MATERIALS SCIENCE AND TECHNOLOGIES

SUPERCONDUCTIVITY

PROPERTIES, APPLICATIONS AND NEW DEVELOPMENTS

MATERIALS SCIENCE AND TECHNOLOGIES

Additional books in this series can be found on Nova's website under the Series tab.

Additional e-books in this series can be found on Nova's website under the e-books tab.

MATERIALS SCIENCE AND TECHNOLOGIES

SUPERCONDUCTIVITY

PROPERTIES, APPLICATIONS AND NEW DEVELOPMENTS

PAULETTE GRANT
EDITOR

nova publishers
New York

Library of Congress Cataloging-in-Publication Data

ISBN: 978-1-63483-907-5

Published by Nova Science Publishers, Inc. † New York

CONTENTS

PREFACE

This book presents research on superconductivity. The first chapter discusses the physical properties of two nonmagnetic noncentrosymmetric superconductors, $LaRhSi_3$ and $LaIrSi_3$/. Chapter two reviews the $SrFe_2As_2$ and $FeTe_{0.8}S_{0.2}$ compounds, which are perspective for developing new superconducting materials on their basis. The possibility of selective excitation of the nonlinear lattice vibrations in high-temperature superconductors formed from layers or thin films opens the new perspectives in the manipulations and control of quantum field in superconducting phase, which is further discussed in chapter three. Chapter four begins with a detailed derivation of the three coupled integral equations associated with the so-called generalized Bose-Einstein condensation formalism of a many-fermion system. Finally, chapter five proposes a microscopic theory of superconductivity based on the spinfluctuation mechanism of pairing.

Chapter 1 - The authors present the physical properties of two nonmagnetic noncentrosymmetric superconductors $LaRhSi_3$ and $LaIrSi_3$ which crystallize in $BaNiSn_3$-type body-centered tetragonal structure (space group *I4mm*) that lacks an inversion symmetry along the crystallographic *c*-axis. The lack of inversion symmetry in crystal structure can spin-split the energy bands and Fermi surface through antisymmetric spin-orbit coupling which in turn may strongly modify the superconducting properties. Investigations of superconducting and normal state properties by various complementary techniques such as heat capacity, magnetic susceptibility, electrical resistivity and muon spin relaxation measurements have revealed a weakly coupled electron-phonon-driven type-I *s*-wave superconductivity in both $LaRhSi_3$ and $LaIrSi_3$. A moderately anisotropic superconducting order parameter is also inferred. The superconducting state properties of $LaRhSi_3$ and $LaIrSi_3$ are described with an emphasis on the analysis of the superconducting state heat capacity data by the α-model of BCS superconductivity.

Chapter 2 - $SrFe_2As_2$ and $FeTe_{0.8}S_{0.2}$ compounds are layered ferriferous compounds, which are perspective for developing new superconducting materials on their basis. For these compounds, the effect of water-induced superconductivity was discovered. This effect occurs due to the diffusion of environmental water vapors into the compound after exposing the virgin samples of compounds to an ambient atmosphere at room temperature. Up to now, there has been no satisfactory explanation of this effect, but its occurrence is generally associated with compression of the compound crystal lattice as the result of water vapor diffusion. The cause of such compression is also not determined. The analysis of a possible processes taking place in the quasi two-dimensional layers of diffused water, was conducted in this work with using of the theory of J. Preparata, who from the position of quantum field theory, presented the description of the processes of water molecules condensation to liquid state in the case of usual water. It is shown that the properties of water in the quasi-two-dimensional structure are significantly different from the properties of usual water. The reasons on which such water can compress a crystalline lattice and induce the arising of superconductivity are also stated. The properties of water diffusing into $SrFe_2As_2$ and $FeTe_{0.8}S_{0.2}$ compound are compared with the properties of nanostructure water in carbon nanotubes (water with a quasi-one-dimensional structure) at low temperatures. The possibility of selective excitation of the nonlinear lattice vibrations in high-temperature superconductors formed from layers or thin films (like in $YBa_2Cu_3O_{6.5}$, MgB_2, $La_{0.7}Sr_{0.3}MnO_3$, graphite layers, etc) opens the new perspectives in the manipulations and control of quantum field in superconducting phase. In many cases the largest critical current is obtained in for thin films or multilayer structures as a function of the constraints of vibration or electron states.

Chapter 3 - The author's note that the anisotropy of electronic and vibrational structures of the materials plays an important role in superconducting critical current or phase transition temperature. In this context the non-linear mechanisms of cooperative interaction between the layer current carriers through the anisotropy of the vibration states of thermostat are examined. One of such mechanism corresponds to the cooperative nonlinear interaction between the cooper pairs belonging to different layers in the single and two-phonon (or vibration) exchanges between the electrons. In this case the constant of cooperative interactions between the electrons increases as a function of temperature in the superconductivity phase. This effect is accompanied by the increasing of order parameter as a function of temperature, after which it begins to decrease till critical temperature. The new ideas of non-linear interaction of carriers with lattice vibration in superconductivity phase of multilayer material are proposing taking into consideration the anisotropy of lattice vibrations and nonlinear aspects of interaction of electrons with lattice vibrations. In the first case the coupling of longitudinal and transversal non-linear oscillation modes in multilayer superconductor materials substantially changes the dispersion law and the Debye frequency of the spectrum of the vibrational modes in quasi-harmonic approximation. As a mapping of this nonlinear coupling, it simulates the temperature increases of the correlation length between the carriers and the coupling integral between the them. The possibilities of the induced generation of coherent vibration modes in the transversal (or longitudinal) geometry of multilayer systems and the control of critical temperature are discussed in the contexts of modern experiments with Raman stimulation of these lattice vibrations in superconducting materials. The second aspect of nonlinear interaction is described by the Hamiltonian function, which takes into consideration two- and single quantum exchanges between the electrons through the lattice vibrations at finite temperature. The Cooper pairs and their coherent behavior are obtained, generalizing the method of elimination of lattice vibration operators for finite temperature. Such a vibration stimulation allows the control of solids in their electronic ground state and the reduced dissipation of direct lattice excitation makes it attractive for applications in functional material control. Many experimental results in this field shown that the electron interaction with non-linear lattice vibration is an important ingredient of the current scenario and can explain superconductivity in novel systems including high temperature superconducting.

Chapter 4 - Begins with a detailed derivation of the three coupled integral equations associated with the so-called generalized Bose-Einstein condensation (GBEC) formalism of a many-fermion system. The formalism can describe superconductors if the fermions are either electrons or holes. Changing the two-fermion interactions allows description of neutral-fermion superfluids (such as liquid ^3He, neutron matter, or ultracold trapped Fermi gases like ^{40}K or ^7Li). In essence, the GBEC formalism accounts not for boson-boson interactions (as BCS theory also ignores) but for boson-fermion (BF) interaction vertices. It leads to two gap-like and a number equation. The three unknowns associated with the model are functions of temperature, coupling and initial fermion number-density, and are: the fermion chemical potential and the BEC number-densities of zero total (or center-of-mass) momentum of two-electron (2e) and two-hole (2h) Cooper pairs (CPs) that continually form and disintegrate within the many-fermion medium as a result of the vertex (two-fermion/one boson) interactions. A detailed sketch follows of how the three equations reduce in the appropriate limits to the five known continuum (as opposed to "spin" or "lattice") statistical models of superconductivity (and by extension to superfluidity), including BCS, BCS-Bose *crossover*, the Friedberg-T.D. Lee and BEC theories. It is also seen that by ignoring either 2h-CPs or 2e-CPs the GBEC cannot give the BCS temperature-dependent gap equation (nor, therefore, the BCS critical temperature). Although explicit mention of individual holes is unnecessary, both kinds of CPs are needed. Finally, practical consequences for 3D superconductors are reviewed which feature critical BEC temperatures T_c enhanced by several orders of magnitude over those of BCS theory, assuming the same interfermion interaction, namely the Cooper-BCS (four-fermion) interaction model corresponding to the assumed BF vertex interaction. These enhancements in T_c result from the GBEC's ability to depart from the perfect 2e/2hCP symmetry to which BCS theory is restricted by construction.

Chapter 5 - Proposes a microscopic theory of superconductivity based on the spin-fluctuation mechanism of pairing. The author's argue that the cuprate superconductors are strongly correlated electronic systems where electronic excitations have to be described by the projected (Hubbard) electron operators in the singly- and doubly occupied Hubbard subbands. Due to the unconventional commutation relations for the Hubbard operators, a specific kinematical interaction of electrons with spin and charge fluctuations emerges. A new energy scale of the order of kinetic energy of electrons W appears in the intraband hopping which is much larger than the antiferromagnetic exchange interaction J induced by the interband hopping. Taking into account this interaction, a microscopic theory of superconductivity in cuprates is presented within the extended Hubbard model where the intersite Coulomb repulsion V and electron-phonon interaction are taken into account. The Dyson equation for the normal and pair Green functions for two Hubbard subbands is derived using Mori-type projection technique. The self-energy is calculated in the self-consistent Born approximation. The author's found the *d*-wave pairing with high-T_c mediated by the kinematical spin-fluctuation interaction. Contributions coming from the realistic in cuprates intersite Coulomb repulsion and electron-phonon interaction turned out to be small since only the angular momentum $l=2$ of interactions gives contributions to the *d*-wave pairing. Isotope effect on T_c induced by electron-phonon interaction is weak at optimal doping and increased at low doping. Superconductivity can be suppressed only for a large intersite Coulomb repulsion $V > W$. The kinematical interaction is absent in the spin-fermion models and is lost in the slave-boson (fermion) models treated in the mean-field approximation.

In: Superconductivity
Editor: Paulette Grant

ISBN: 978-1-63483-907-5

Chapter 1

TYPE-I SUPERCONDUCTIVITY IN NONCENTROSYMMETRIC $LaRhSi_3$ AND $LaIrSi_3$

***V. K. Anand*[1,2,*], *D. T. Adroja*[1,3,†] *and A. M. Strydom*[3]**
[1]ISIS Facility, Rutherford Appleton Laboratory, Chilton, Didcot, Oxon, United Kingdom
[2]Helmholtz-Zentrum Berlin für Materialien und Energie GmbH, Berlin, Germany
[3]Highly Correlated Matter Research Group, Physics Department, University of Johannesburg, Johannesburg, South Africa

Abstract

In this chapter we present the physical properties of two nonmagnetic noncentrosymmetric superconductors $LaRhSi_3$ and $LaIrSi_3$ which crystallize in $BaNiSn_3$-type body-centered tetragonal structure (space group $I4\,mm$) that lacks an inversion symmetry along the crystallographic c-axis. The lack of inversion symmetry in crystal structure can spin-split the energy bands and Fermi surface through antisymmetric spin-orbit coupling which in turn may strongly modify the superconducting properties. Investigations of superconducting and normal state properties by various complementary techniques such as heat capacity, magnetic susceptibility, electrical resistivity and muon spin relaxation measurements have revealed a weakly coupled electron-phonon-driven type-I s-wave superconductivity in both $LaRhSi_3$ and $LaIrSi_3$. A moderately anisotropic superconducting order parameter is also inferred. The superconducting state properties of $LaRhSi_3$ and $LaIrSi_3$ are described with an emphasis on the analysis of the superconducting state heat capacity data by the α-model of BCS superconductivity.

1. Introduction

The noncentrosymmetric superconductors (NCSs) which lack an inversion symmetry in their crystal structure present exotic superconducting state properties due to the substantial

*E-mail address: vivekkranand@gmail.com
†E-mail address: devashibhai.adroja@stfc.ac.uk

modification of energy bands and Fermi surface by the antisymmetric spin-orbit coupling (ASOC). Because of the lack of spatial inversion symmetry the spin-orbit coupling acquires an antisymmetric component with respect to the wave-vector **k** which is referred as ASOC. The ASOC removes the spin degeneracy of the conduction band electrons, splitting the spin-up and spin-down energy bands, leading to an admixture of spin-singlet and spin-triplet states, which makes the noncentrosymmetric superconductivity distinct from the conventional BCS superconductivity. A schematic representation of Fermi surfaces of NCSs and conventional superconductors are shown in Fig. 1. The spin-split Fermi surfaces for spin-up and spin-down bands have very similar topology but different volumes. Due to ASOC the spins of the conduction electrons (subject to an effective magnetic field related to electron momentum and non-uniform lattice potential) would rotate around the Fermi surface, clockwise or anticlockwise depending on the spin-states (up or down).

The noncentrosymmetric superconductors captured much attention after the finding of anomalous superconductivity in $CePt_3Si$ by Bauer et al. in 2004 [1]. This compound has a primitivie tetragonal symmetry ($P4mm$) that lacks an inversion symmetry along the crystallographic c-axis which is the key for the anomalous superconducting properties [1–3]. The advent of superconductivity in noncentrosymmetric $CePt_3Si$ triggered momentum for a hunt of new NCSs leading to the discovery of many NCSs among which the Ce-based NCSs $CeRhSi_3$ [4, 5], $CeIrSi_3$ [6], $CeCoGe_3$ [7, 8] and $CeIrGe_3$ [9] belonging to the class of heavy fermion superconductors like $CePt_3Si$ present intriguing superconducting properties [3].

The nature of the superconducting ground state is determined by the symmetry of the superconducting order parameter. The two fundamental symmetries for the Cooper pair

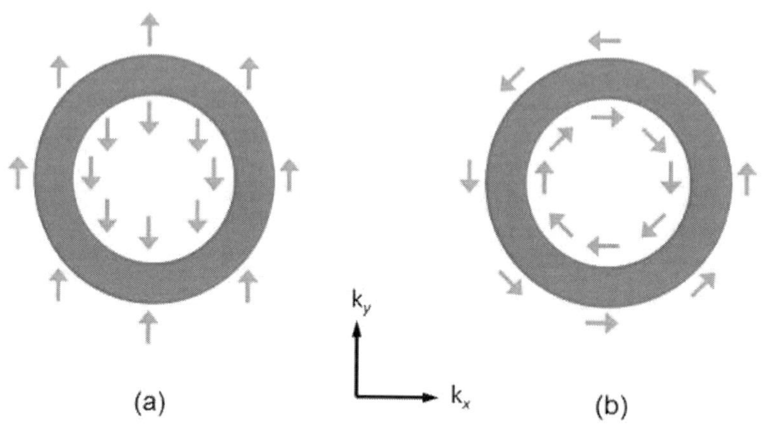

Figure 1. (Color online) A schematic representation of Fermi surface in momentum-space for (a) a conventional BCS superconductor, and (b) spin-splitting with a **k**-dependent Rashba-type antisymmetric spin-orbit coupling (ASOC) in a noncentrosymmetric superconductor having nonuniform lattice potential along z-axis. The arrows indicate spins of the conduction electrons for the spin-up (outer ring) and spin-down (inner ring) bands. The ASOC causes the spins (subject to an effective magnetic field) to rotate around the Fermi surface, clockwise or anticlockwise depending on the spin-states (Ref. [10]).

formation are the time reversal symmetry and the inversion symmetry that determine the parity states (even-parity spin-singlet and odd-parity spin-triplet). The absence of either of the two symmetries can lead to the spin-splitting of the Fermi surface (spin-up and spin-down energy bands) that strongly influences the Cooper pairing. The lack of inversion symmetry in crystal structure leads to a nonuniform lattice potential which causes an antisymmetric spin-orbit coupling, thereby the spin-up and spin-down energy bands are no longer degenerate, and the spin orientation becomes **k**-dependent [see Fig. 1]. Thus, with a spin-split Fermi surface the electrons participating in the formation of superconducting Cooper pairs belong to two different kinds of Fermi surfaces, separated in energy by about 10–1000 K. The Cooper pair formation is allowed only within each Fermi surface, i.e., by the electrons belonging to only one kind of Fermi surface. A simple p-wave pairing (odd parity) is prohibited, as the formation of such Cooper pairs would require electrons from both the spin-split Fermi surfaces which is not possible because of the large energy separation between the two Fermi surfaces, much larger than the superconducting energy gap which is usually only a few Kelvin. With these constraints and because of the restriction on the spin orientation, the spin and orbital parts of the Cooper pairs of NCSs cannot be treated independently. The Cooper pair formation in NCSs is thus substantially different from the conventional superconductors where the electrons forming Cooper pairs belong to the single Fermi surface. The Cooper pair wave function of a conventional BCS superconductor has even parity characterized by a symmetric orbital state and an antisymmetric spin state, usually referred as 's-wave spin-singlet'. In conventional superconductors only the gauge symmetry is broken. In addition in NCSs the inversion symmetry is also broken, therefore the Cooper pair wave function of NCSs cannot be treated as an even-parity spin-singlet or odd-parity spin-triplet, but instead forms a mixed parity two component order parameter consisting of mixed spin-singlet and spin-triplet states. The relative magnitude of singlet and triplet components depends on the strength of the ASOC. Thus from theoretical viewpoints the structural inversion symmetry and ASOC have profound effects on the superconductivity in NCSs [3, 11–15].

The ASOC dictated exotic superconductivity has been observed in the Ce-based heavy fermion NCSs $CePt_3Si$, $CeRhSi_3$, $CeIrSi_3$, $CeCoGe_3$ and $CeIrGe_3$ [1–9]. Among these only $CePt_3Si$ shows superconductivity at ambient pressure and others superconduct only under pressure. Some of the striking features of the anomalous superconductivity in $CePt_3Si$ are: a) observation of a peak in NMR relaxation suggesting a nodeless energy gap spin-singlet superconductivity, b) observation of power-law feature in magnetic penetration depth and thermal conductivity suggesting the existence of a line node in the superconducting gap, and c) the finding of upper critical magnetic field exceeding the Pauli-limiting upper critical field which could not be understood in an s-wave spin-singlet picture, thus favoring the spin-triplet pairing, which, however is prohibited by the absence of inversion symmetry [2, 3]. These contradictory findings are accounted by a two-component order parameter consisting of mixed spin-singlet and spin-triplet states. The Ref. [3] presents a good review of the superconductivity in these Ce-based heavy fermion NCSs and the theoretical developments.

In this chapter we present the physical properties of two nonmagnetic NCSs $LaRhSi_3$ [16] and $LaIrSi_3$ [17] which do not contain magnetic moment carrying atoms like Ce-based NCSs in which the presence of strong electronic correlations make it difficult to understand the role of crystallographic inversion symmetry and ASOC in determining the supercon-

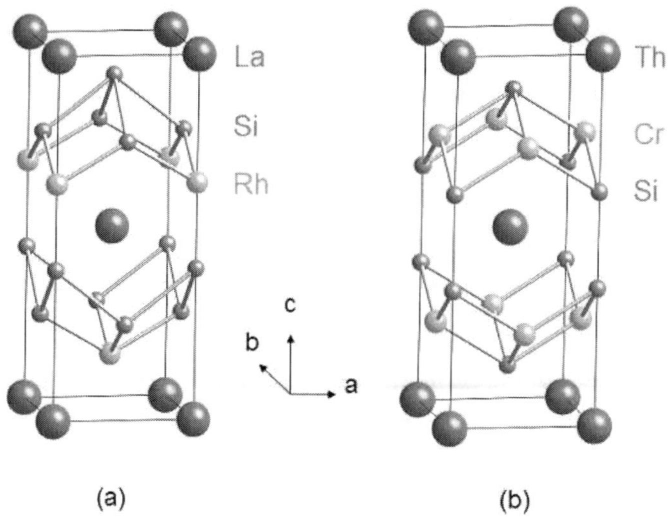

Figure 2. (Color online) Comparison of (a) noncentrosymmetric $BaNiSn_3$-type body-centered tetragonal structure (space group $I4\,mm$) of $LaRhSi_3$ and (b) centrosymmetric $\mathrm{ThCr_2Si_2}$-type body-centered tetragonal crystal structure (space group $I4/mmm$).

ducting state properties. In the following sections we give an overview of the crystal structure and discuss the superconducting state properties through magnetic susceptibility $\chi(T)$, isothermal magnetization $M(H)$, electrical resistivity $\rho(T)$ and heat capacity $C_{\rm p}(T)$ of the two compounds where T denotes temperature and H the magnetic field. A summary of the chapter is given at the end.

2. Crystal Structure

The refinements of powder x-ray diffraction data have revealed that both $LaRhSi_3$ and $LaIrSi_3$ crystallize in $BaNiSn_3$-type body-centered tetragonal (bct) structure (space group $I4\,mm$) with lattice parameters a = 4.2693(4) Å and c = 9.8292(9) Å for $LaRhSi_3$ and a = 4.2784(3) Å and c = 9.8308(7) Å for $LaIrSi_3$. The lack of inversion symmetry in $BaNiSn_3$-type tetragonal structure comes from the absence of a mirror plane symmetry along the c-axis. This structure belongs to the same irreducible representation point group C_{4v} as does $CePt_3Si$ which is favored for Rashba-type ASOC. The $BaNiSn_3$ bct structure is illustrated in Fig. 2 along with the common bct $ThCr_2Si_2$ structure (space group $I4/mmm$) having an inversion symmetry.

The $BaNiSn_3$ structure shares some common feature with $ThCr_2Si_2$ structure, both are ternary derivatives of $BaAl_4$-type structure and are layered [18]. The A (Ba, Th) atoms occupy the same crystallographic site in both the structures (Wykoff position 2a) though the positions of T (Ni, Cr) and X (Sn, Si) atoms are different. They both have body-centered tetragonal sublatttice of A atoms. The T atoms occupy sites with Wyckoff positions 2a in $BaNiSn_3$ structure and 4d in $ThCr_2Si_2$ structure. In both the structures the T atoms form square sublattice in ab-plane however with a relative ab-plane rotation of 45°. Another difference among these structures is that in $BaNiSn_3$ structure the X atoms occupy two different sites (Wykoff positions 2a & 4b) whereas in $ThCr_2Si_2$ structure the X atoms occupy only one crystallographic site (Wykoff position 4e). Consequently they have very different stacking order of the layers of T and X atoms along the c-axis. The comparison of two structures in Fig. 2 clearly reflects the structural similarities and differences. While the $ThCr_2Si_2$ structure is symmetric about the R plane, this symmetry is missing in $BaNiSn_3$ structure. Thus $ThCr_2Si_2$ structure has an inversion symmetry and is referred as centrosymmetric. However, the loss of mirror plane in $BaNiSn_3$ structure reflects an absence of inversion symmetry and the structure is noncentrosymmetric.

3. $LaRhSi_3$

3.1. Overview of Superconductivity

A sharp superconducting transition at $T_c = 2.16(8)$ is evident from the $\chi(T)$, $\rho(T)$ and $C_p(T)$ data [Fig. 3] of $LaRhSi_3$ [16]. Large Meissner signal in zero-field-cooled (ZFC) $\chi(T)$, zero resistance state in $\rho(T)$ and sharp jump in the heat capacity at T_c demonstrate the bulk nature of the superconductivity in $LaRhSi_3$. The ZFC volume susceptibility is around $4\pi\chi_V \sim -1$ indicating 100% superconducting phase fraction. The isothermal $M(H)$ [Fig. 3(b)] at 0.5 K which is well below T_c exhibits reentrant diamagnetism for decreasing H cycle (while traversing from normal state to superconducting state) after the superconductivity is destroyed by increasing H, which suggests a type-I superconductivity in this compound. However, the $M(H)$ of $LaRhSi_3$ does not follow the ideal step transition at critical field expected for a type-I superconductor. This type of transition width may arise from the nonzero demagnetization factor due to the geometrical shape of the sample. The demagnetization factor drives the superconducting system to an intermediate state whereby the macroscopic domains of both normal and superconducting phases coexist. Such an intermediate state has also been observed in type-I superconductor LaNiSn [19].

The $C_p(T)$ data measured in different magnetic fields [Fig. 4(a)] show the expected decrease in T_c with the increasing H on account of closing of superconducting energy gap by field. A field of 15.0 mT suppresses the superconductivity to a temperature below 0.4 K. On the other hand the superconducting jump in the heat capacity ΔC_p at T_c increases first at 2.5 mT and then decreases at further higher fields. This indicates a first order transition in $H = 2.5$ mT. A change in the nature of the transition from second order to first order under field further supports the type-I superconductivity inferred from $M(H)$.

3.2. Normal State Properties

The normal state low-T $C_p(T)$ data are well described by $C_p(T) = \gamma_n T + \beta T^3$ [Fig. 4(b)] yielding the normal state Sommerfeld electronic heat capacity coefficient $\gamma_n = 6.04(2)$ mJ/mol K^2 and $\beta = 0.214(5)$ mJ/mol K^4. The coefficient β gives the Debye temperature $\Theta_D = (12\pi^4 Rr/5\beta)^{1/3} = 357(3)$ K (where R is the molar gas constant and r is number of atoms per formula units) and γ_n gives the density of states (DOS) at the Fermi level $\mathcal{D}(E_F) = 3\gamma_n/\pi^2 k_B^2 = 2.56(1)$ states/eV f.u. for both spin directions. The bare DOS $\mathcal{D}_{band}(E_F) = \mathcal{D}(E_F)/(1 + \lambda_{e-ph})$ where λ_{e-ph} is the electron-phonon cou-

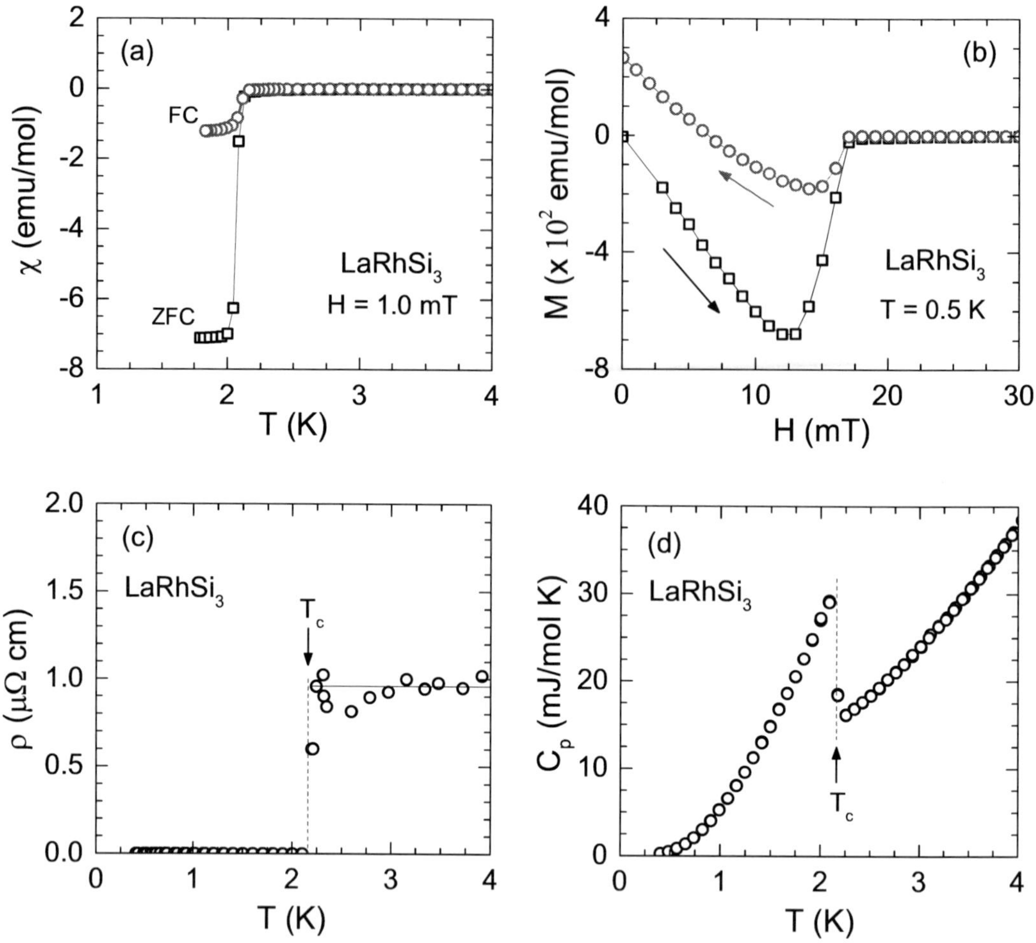

Figure 3. (Color online) (a) Zero-field-cooled (ZFC) and field-cooled (FC) magnetic susceptibility χ of $LaRhSi_3$ as a function of temperature T measured in an applied magnetic field $H = 1.0$ mT. (b) Isothermal magnetization $M(H)$ at $T = 0.5$ K. Arrows indicate the directions for the magnetic field cycle. (c) Electrical resistivity $\rho(T)$ measured in $H = 0$. (d) Heat capacity $C_{\rm p}(T)$ measured in $H = 0$. The arrows in (c) and (d) mark the superconducting transition temperature $T_{\rm c}$. Solid and dotted lines are the guides to eye (Ref. [16]).

pling constant. For a superconductor, McMillan's theory [20] relates $\lambda_{\rm e-ph}$ to $\Theta_{\rm D}$ and $T_{\rm c}$. Accordingly for $LaRhSi_3$ $\lambda_{\rm e-ph} \approx 0.50$ and $\mathcal{D}_{\rm band}(E_{\rm F}) = 1.71(1)$ states/eV f.u. for both spin directions. The $\lambda_{\rm e-ph}$ gives a measure of attractive part of Cooper pair bonding, and the small value of 0.50 indicates a weak-coupling superconductivity in $\rm LaRhSi_3$ as is expected for phonon-mediated superconductivity. The effective mass of the quasi-particle $m^* = (1+\lambda_{\rm e-ph})\, m^*_{\rm band} = 1.5\, m_{\rm e}$ where we assume the effective band mass $m^*_{\rm band} = m_{\rm e}$, $m_{\rm e}$ being the mass of a free electron. This estimate of m^* is in good agreement with the dHvA measurements ($m^* \sim 1.6\, m_e$) [21].

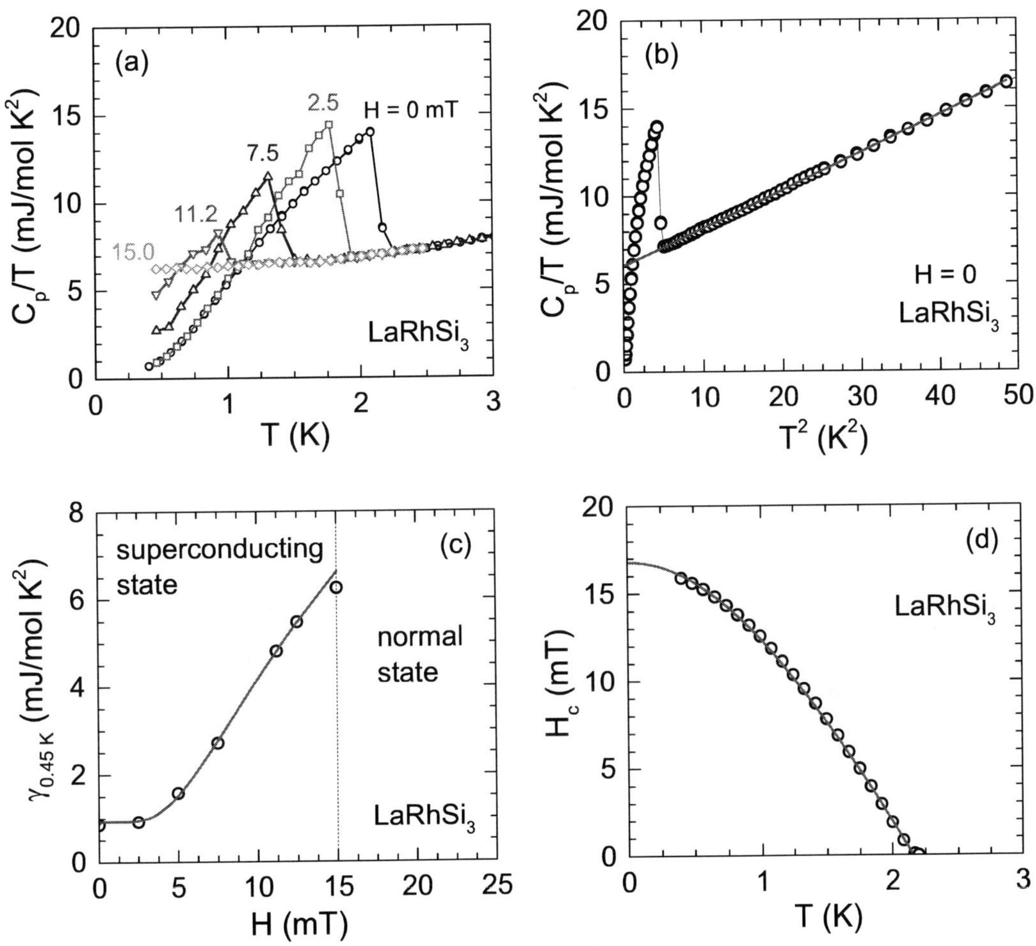

Figure 4. (Color online) (a) Heat capacity C_{p} of $\mathrm{LaRhSi_3}$ as a function of temperature T measured in different magnetic fields $0 \leq H \leq 15.0$ mT, plotted as C_{p}/T versus T. (b) Zero-field C_{p}/T versus T^2. The solid line is a linear fit according to $C_{\mathrm{p}}(T)/T = \gamma_{\mathrm{n}} + \beta T^2$ for 2.3 K $\leq T \leq 7$ K. (c) Superconducting state $\gamma \equiv C_{\mathrm{p}}/T$ at $T = 0.45$ K as a function of H. The solid red curve represents an exponential evolution of γ with H, $\gamma(H) \sim \exp(-H^*/H)$. (d) Thermodynamic critical field $H_c(T)$ determined from the experimentally derived electronic heat capacity $C_{\mathrm{e}}(T)$. The solid red curve is the theoretical prediction of the α-model for $\alpha = 1.67$ (Ref. [16]).

The $\mathcal{D}(E_{\mathrm{F}})$ is related to Fermi velocity v_{F} by [22]

$$v_{\mathrm{F}} = \frac{\pi^2\hbar^3}{m^{*2}V_{\mathrm{f.u.}}}\mathcal{D}(E_{\mathrm{F}}) \tag{1}$$

where $V_{\mathrm{f.u.}}$ is volume per formula unit. $V_{\mathrm{f.u.}} = V_{\mathrm{cell}}/Z$ where V_{cell} is unit cell volume and Z the number of formula units per unit cell; $Z = 2$ for $\mathrm{LaRhSi_3}$. Eq. (1) gives $v_{\mathrm{F}} = 1.11 \times 10^8\ \mathrm{cm/s}$. The mean free path $\ell = v_{\mathrm{F}}\tau$ with the mean free scattering time $\tau = m^*/ne^2\rho_0$ where n is the conduction carrier density and ρ_0 the residual resistivity before entering the

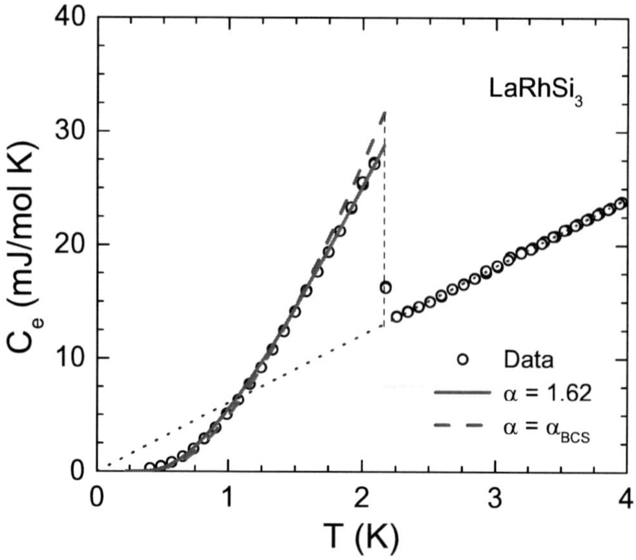

Figure 5. (Color online) Electronic contribution C_e to zero field heat capacity as a function of temperature T for LaRhSi$_3$. The solid red curve is the theoretical prediction of the α-model for $\alpha = \Delta(0)/k_BT_c = 1.62$. The BCS prediction for $\alpha_{BCS} \approx 1.764$ is also shown for comparison. The dotted blue curve represents the normal state $\gamma_n T$.

superconducting state. For a spherical Fermi surface $n = m^{*3}v_F^3/3\pi^2\hbar^3$ [22]. Thus

$$\ell = 3\pi^2\left(\frac{\hbar}{e^2\rho_0}\right)\left(\frac{\hbar}{m^*v_F}\right)^2. \tag{2}$$

For LaRhSi$_3$ $\rho_0 = 1\ \mu\Omega\,\text{cm}$ therefore $\ell = 59.2\ \text{nm}$ which is about 60 c-axis lattice spacings.

3.3. Superconducting State Properties

The $C_p(T)$ data [Fig. 3(d)] show sharp jump at the superconducting transition. In order to estimate the jump in heat capacity at T_c we separated out the electronic contribution $C_e(T)$ to heat capacity as shown in Fig. 5 which was estimated by subtracting off the phonon contribution as $C_e(T) = C_p(T) - \beta T^3$. A jump of $\Delta C_e = 7.3(1)$ mJ/mol K implies $\Delta C_e/\gamma_n T_c = 1.21(1)$. This value of $\Delta C_e/\gamma_n T_c$ is smaller than the BCS value of 1.43 [23]. One possible cause for the reduction in $\Delta C_e/\gamma_n T_c$ can be the presence of a small nonsuperconducting phase in the sample causing a residual heat capacity thereby the estimated γ_n is higher than the true γ which being in denominator yields a reduced value of $\Delta C_e/\gamma_n T_c$. In this case one can adjust γ_n in a way that it produces the BCS value 1.43. This is the approach that was followed in Ref. [16] in order to analyze the superconducting state heat capacity data of LaRhSi$_3$ by the generalized weak coupling BCS dataset of Mühlschlegel [24].

Another possible cause for the reduction in $\Delta C_e/\gamma_n T_c$ could be the anisotropy of order parameter in single-band picture or from multiband nature of superconductivity [25]. Considering that the jump in $C_e(T)$ at T_c is very sharp and the superconducting volume fraction from $\chi(T)$ is nearly 100% one would expect almost no residual γ. Therefore the reduction in $\Delta C_e/\gamma_n T_c$ could be intrinsic possibly reflecting the anisotropic superconducting energy gap (order parameter) in momentum space [25]. If this is the case one can analyze the superconducting state heat capacity data by the single-band α-model of BCS superconductivity [25–27] which is formulated for $\Delta C_e/\gamma_n T_c \neq 1.43$.

The α-model of BCS superconductivity has recently been reviewed by Johnston [25] and has been successfully applied for the analysis of the superconducting state heat capacity data of a few systems such as $CaPd_2As_2$ [28] and $CaPd_2Ge_2$ [29] with reduced $\Delta C_e/\gamma_n T_c$. In the so-called α-model the order parameter and London penetration depth are calculated using $\alpha_{BCS} \equiv \Delta(0)/k_B T_c = 1.764$, however a variable α is used for fitting the superconducting state thermodynamic data [25, 27]. The value of α is determined from [25]

$$\frac{\Delta C_e(T_c)}{\gamma_n T_c} = 1.426 \left(\frac{\alpha}{\alpha_{BCS}}\right)^2. \tag{3}$$

For $LaRhSi_3$ with $\Delta C_e/\gamma_n T_c = 1.21(1)$ we obtain $\alpha = 1.62(1)$ which is smaller than $\alpha_{BCS} = 1.764$. The α-model $C_e(T)$ calculated for $\alpha = 1.62$ is shown by the solid (red) curve in Fig. 5. For comparison the calculation for $\alpha = \alpha_{BCS}$ is also shown by the dashed (olive) curve. It is evident from Fig. 5 that BCS theory prediction for α_{BCS} fails to describe the observed $C_e(T)$. However, a good agreement is observed between the $C_e(T)$ and the α-model prediction. Thus the analysis of $C_e(T)$ by α-model suggests an anisotropic energy gap in $LaRhSi_3$.

Further support for anisotropic energy gap comes from the field dependence of γ. For an isotropic energy gap s-wave superconductor γ is linear in H. In contrast, the $\gamma(H) = C(T)_H/T$ at $T = 0.45$ K follows $\gamma(H) \sim \exp(-H^*/H)$ with $H^* \approx 17$ mT [see Fig. 4(c)]. The nonlinear behavior of $\gamma(H)$ clearly indicates a non-isotropic order parameter in $LaRhSi_3$. This exponential evolution of γ and hence the nonsuperconducting density of states with H is very peculiar and might possibly have some relation with ASOC which is reinforced by magnetic field manifesting the strongly field-dependent anisotropic order parameter.

The thermodynamic critical field H_c in α-model can be estimated by [25]

$$\frac{H_c(0)}{(\gamma_{nV} T_c^2)^{1/2}} = \sqrt{\frac{6}{\pi}}\, \alpha \approx 1.382\, \alpha, \tag{4}$$

where the Sommerfeld coefficient per unit volume $\gamma_{nV} = 1120\ \mathrm{erg/cm^3\,K^2}$ for $LaRhSi_3$. Accordingly for $\alpha = 1.62$, $H_c(0) = 16.2$ mT. The $H_c(T)$ can be estimated from $C_e(T)$ using [23, 30]

$$H_c^2(T) = 8\pi \int_T^{T_c} [S_{en}(T') - S_{es}(T')]dT'. \tag{5}$$

Here S_{en} and S_{es} are the electronic entropy of normal and superconducting states, respectively, with $S_e(T') = \int_0^{T'} [C_e(T'')/T'']dT''$. The experimentally derived $H_c(T)$ from $C_e(T)$ (at $H = 0$) is shown in Fig. 4(d). The $H_c(T)$ calculated according to α-model is shown by the solid (red) curve in Fig. 4(d). The calculated $H_c(T)$ for $\alpha = 1.62$ is found to

Table 1. Measured and derived superconducting and relevant normal state parameters for noncentrosymmetric superconductors $LaRhSi_3$ [16] and $LaIrSi_3$ [17]. T_c: bulk superconducting transition temperature; γ_n: Sommerfeld electronic heat capacity coefficient in normal-state; $\mathcal{D}(E_F)$: density of states at the Fermi energy for both spin directions determined from γ_n; Θ_D: Debye temperature from fitting heat capacity data; λ_{e-ph}: electron-phonon coupling constant; ΔC_e: heat capacity jump at T_c; $\Delta(0)$: superconducting order parameter; H_c: thermodynamic critical magnetic field; κ_{GL}: Ginzburg-Landau parameter; $\xi(0)$: Ginzburg-Landau coherence length; ξ_0: BCS superconducting coherence length; ℓ: electronic mean-free path at low T; $\lambda_L(0)$: London penetration depth; λ_{eff}: effective magnetic penetration depth. *Note*: Some of the parameters for $LaRhSi_3$ have been recalculated for $m^* = (1 + \lambda_{e-ph})\, m_e$ and for α-model of BCS superconductivity, therefore they are different from the values reported in Ref. [16]

Property	$LaRhSi_3$	$LaIrSi_3$
T_c (K)	2.16(8)	0.72(1)
γ_n (mJ/mol K^2)	6.04(2)	4.60(2)
$\mathcal{D}(E_F)$ (states/eV f.u.)	2.56(1)	1.95(1)
Θ_D (K)	357(3)	385(8)
λ_{e-ph}	0.50	0.41
ΔC_e (mJ/mol K)	7.3(1)	3.6(1)
$\Delta C_e/\gamma_n T_c$	1.21(1)	1.09(3)
$\alpha \equiv \Delta(0)/k_B T_c$ (from $\Delta C_e/\gamma_n T_c$)	1.62(1)	1.54(2)
$\Delta(0)/k_B$ (K)	3.50	1.11
$H_c(T=0)$ (mT)	16.7(1)	6.10(3)
κ_{GL}	0.25	0.55
$\xi(0)$ (nm)	309	373
ξ_0 (nm)	769	2081
ℓ [for $m^* = (1 + \lambda_{e-ph})\, m_e$] (nm)	59.2	33.7
$\lambda_L(0)$ (clean limit) (nm)	20.6	25.9
$\lambda_{eff}(0)$ (dirty limit) (nm)	77	205

be little lower in magnitude than the experimental $H_c(T)$ data, instead a value of $\alpha = 1.67$ reproduces the data with $H_c(0) = 16.7$ mT.

The superconducting parameter London penetration depth $\lambda_L(0)$ can be estimated from v_F using the relation [23]

$$\lambda_L(0)^2 = \frac{m^* c^2}{4\pi n e^2} = \frac{3\pi c^2 \hbar^3}{4 m^{*2} e^2 v_F^3}, \tag{6}$$

which gives $\lambda_L(0) = 20.6$ nm. Within the α-model, the BCS coherence length ξ_0 is given by [25]

$$\xi_0 = \frac{\hbar v_F}{\pi \Delta(0)} = \left(\frac{1}{\pi\alpha}\right) \frac{\hbar v_F}{k_B T_c}, \tag{7}$$

which for $\alpha = 1.62$ yields $\xi_0 = 769$ nm. The ξ_0 is much larger than the $\ell = 59.2$ nm,

$\ell/\xi_0 \approx 0.077 \ll 1$ which indicates that $LaRhSi_3$ is a dirty-limit superconductor. For the dirty-limit case, the Ginzburg-Landau parameter $\kappa_{GL} = 0.715\,\lambda_L(0)/\ell = 0.25$ for $LaRhSi_3$. The $\kappa_{GL} = 0.25$ is smaller than the value $1/\sqrt{2}$ classifying $LaRhSi_3$ as a type-I superconductor. For the dirty-limit case, the effective magnetic penetration depth $\lambda_{\rm eff}$ is given by [23]

$$\lambda_{\rm eff}(0) = \lambda_L(0)\sqrt{1+\frac{\xi_0}{\ell}} \qquad \text{(dirty limit)}. \tag{8}$$

yielding $\lambda_{\rm eff}(0) = 77.2$ nm. The Ginzburg-Landau coherence length $\xi(0)$ can be estimated using the relation $\kappa_{GL} = \lambda_{\rm eff}(0)/\xi(0)$ yielding $\xi(0) = 308.8$ nm. The superconducting parameters of $LaRhSi_3$ are listed in Table 1 together with those of $LaIrSi_3$.

4. $LaIrSi_3$

The zero-field heat capacity data of $LaIrSi_3$ [Fig. 6(a)] show a well defined superconducting transition at $T_c = 0.72(1)$ K [17]. Like $LaRhSi_3$, the jump in the heat capacity at the superconducting transition in $LaIrSi_3$ too is significantly reduced compared to BCS value. The jump in $C_e(T)$ at $T = T_c$ yields $\Delta C_e/\gamma_n T_c = 1.09(3)$ where $\gamma_n = 4.60(2)$ mJ/mol K^2 is estimated from the normal state heat capacity. The value 1.09(3) is much smaller than the BCS expected value of 1.43. Considering the sharp feature of superconducting jump ΔC_e and bulk nature of superconductivity the reduced value of $\Delta C_e/\gamma_n T_c$ seems to be intrinsic. Therefore like $LaRhSi_3$ the order parameter in $LaIrSi_3$ could also be anisotropic. The superconducting state $C_e(T)$ data could be reasonably described by single-band α-model of BCS superconductivity [27, 25]. The $\Delta C_e/\gamma_n T_c = 1.09(3)$ corresponds to $\alpha = 1.54(2)$ [Eq. (3)] which is significantly smaller than the $\alpha_{BCS} = 1.764$. The α-model prediction for $\alpha = 1.54$ is shown by the solid (red) curve in Fig. 6(c) and compared with that for α_{BCS} [dashed (olive) curve]. The good agreement between $C_e(T)$ and the α-model prediction supports the applicability of α-model indicating an anisotropic energy gap in $LaIrSi_3$.

The $\rho(T)$ data on the other hand [Fig. 6(d)] shows a higher $T_{c\,\rm onset} \approx 1.6$ K and zero resistivity state at $T_{c\,0} \approx 1.3$ K with $T_{c\,\rm mid} = 1.45$ K. This could possibly be due to a filamentary nonbulk superconductivity setting in at a higher T_c than the bulk superconducting T_c. Further it is seen from the $\rho(T)$ data measured in applied field [Fig. 6(d)] that the superconductivity survives even at 0.40 T which is very different from the behavior of $C_p(T)$ under field [Fig. 6(b)] whereby a field of 0.07 T suppresses the T_c to a temperature below 0.35 K. Thus the suppression of T_c with H is very different for $\rho(T,H)$ and $C_p(T,H)$ data, apparently due to the filamentary superconductivity sensed in $\rho(T,H)$ measurement.

The κ_{GL} estimated for $LaIrSi_3$ is 0.55 ($< 1/\sqrt{2}$) suggesting a type-I behavior. Filamentary or surface superconductivity is not very common in a type-I superconductor unless the κ_{GL} satisfies the condition $1/2.39 < \kappa_{GL} < 1/\sqrt{2}$ [31, 32] which is satisfied for the present case. Therefore $LaIrSi_3$ can show surface superconductivity. However, the observed critical field associated with the surface superconductivity does not comply with the predicted $H_{c3} = 2.39\,\kappa H_c$ for type-I superconductor [31, 32], which for $LaIrSi_3$ with $H_c(0) = 6.10(3)$ mT gives $H_{c3} \approx 8.0$ mT, much smaller than the H_{c3} inferred from $\rho(T,H)$ [17]. This kind of high critical field in $\rho(T,H)$ measurement than $C_p(T,H)$ has also been observed in type-I superconductor $LaPdSi_3$ [33] as well as in $LaRhSi_3$ [34].

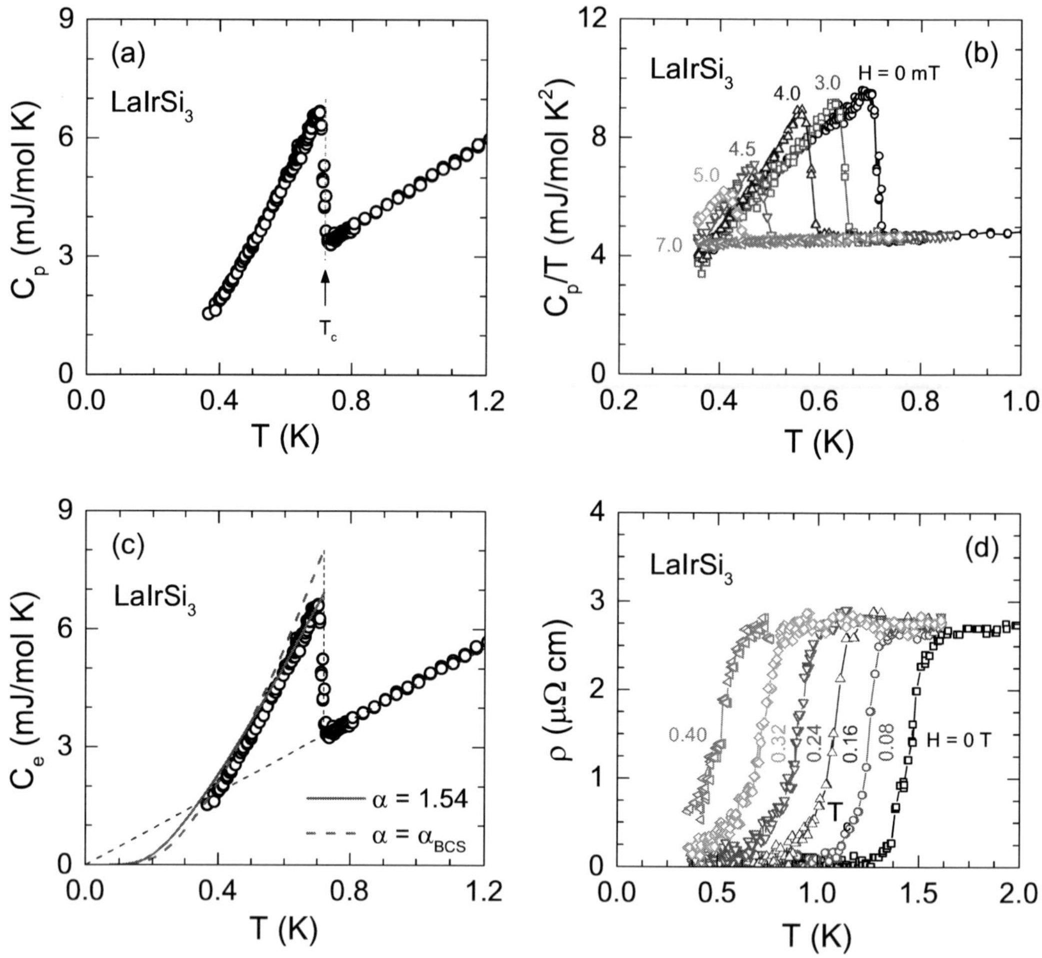

Figure 6. (Color online) (a) Zero field heat capacity $C_{\rm p}$ of $LaIrSi_3$ as a function of temperature T. (b) $C_{\rm p}(T)$ data at different applied magnetic fields $0 \leq H \leq 7.0$ mT, plotted as $C_{\rm p}/T$ versus T. (c) Electronic contribution to zero field heat capacity $C_{\rm e}(T)$. The solid red curve is the theoretical prediction of the α-model for $\alpha = \Delta(0)/k_{\rm B}T_{\rm c} = 1.54$. The BCS prediction for $\alpha_{\rm BCS} \approx 1.764$ is also shown for comparison. The dotted blue curve represents the normal state $\gamma_{\rm n}T$. (d) Electrical resistivity $\rho(T)$ measured in $0 \leq H \leq 0.40$ T (Ref. [17]).

Kimura et al. [34] suggested that this behavior may be a common feature in NCSs however a theoretical verification is lacking at present.

The $\gamma_{\rm n} = 4.60(2)$ mJ/mol K^2 corresponds to $\mathcal{D}(E_{\rm F}) = 1.95(1)$ states/eV f.u. for both spin directions and $\mathcal{D}_{\rm band}(E_{\rm F}) = 1.38(1)$ states/eV f.u. for both spin directions with $\lambda_{\rm e-ph} = 0.41$ estimated from $\Theta_{\rm D} = 385(8)$ K and $T_{\rm c} = 0.72(1)$ K. The $v_{\rm F} = 9.49 \times 10^7$ cm/s [Eq. (1)] is obtained assuming $m^* = (1+\lambda_{\rm e-ph})\, m^*_{\rm band} = 1.41\, m_{\rm e}$. The $v_{\rm F}$ and $\rho_0 = 2.7\ \mu\Omega$ cm gives $\ell = 33.7$ nm [Eq. (2)], and $\lambda_{\rm L}(0) = 25.9$ nm [Eq. (6)]. Using α-model Eq. (7) the ξ_0 for $\alpha = 1.54$ is found to be 2081 nm. The values of ℓ and ξ_0 indicate dirty-limit superconductivity in $LaIrSi_3$ ($\ell/\xi_0 \approx 0.016 \ll 1$). The $\lambda_{\rm eff}(0) = 205$ nm is estimated from Eq. (8). The $\kappa_{\rm GL}$ and $\lambda_{\rm eff}(0)$ yields $\xi(0) = 373$ nm. These parameters for $LaIrSi_3$ are listed in Table 1.

5. Summary

In this chapter we presented an overview of superconductivity in noncentrosymmetric superconductors $LaRhSi_3$ and $LaIrSi_3$, both of which are unencumbered by magnetic elements and magnetic cooperative phenomena. They both exhibit dirty-limit weak-coupling electron-phonon-driven type-I s-wave superconductivity with moderately anisotropic order parameter in momentum space, well accounted by single-band α-model of BCS superconductivity. The type-I behavior and s-wave singlet pairing with preserved time-reversal symmetry have been further confirmed by muon spin relaxation and rotation (μSR) measurements on both $LaRhSi_3$ [16] and $LaIrSi_3$ [17]. Though the single-band α-model of BCS superconductivity reasonably accounts for the observed superconducting properties, due to the presence of spin-split energy bands a multiband (two bands) superconductivity is also tenable. The band-structure calculations and dHvA effect studies have clearly shown the large splitting of Fermi surfaces on account of ASOC in both $LaRhSi_3$ [21] and $LaIrSi_3$ [35]. Therefore an exotic superconductivity seems to be a straightforward conjecture. In contrast, despite the presence of spin-split energy bands and Fermi surfaces no noticeable effect of ASOC is observed in superconducting properties of $LaRhSi_3$ and $LaIrSi_3$ except an unusual field evolution of γ in the case of $LaRhSi_3$. Thus a need of additional control parameter (besides ASOC) is anticipated for the appearance of ASOC dictated anomalous superconductivity in NCSs.

Acknowledgments

Work at RAL was supported by CMPC-STFC grant number CMPC-09108. AMS thanks the URC of UJ and SA-NRF (93549).

References

[1] E. Bauer, G. Hilscher, H. Michor, C. Paul, E. W. Scheidt, A. Gribanov, Y. Seropegin, H. Noel, M. Sigrist, and P. Rogl, *Phys. Rev. Lett.* 92, 027003 (2004).

[2] E. Bauer, H. Kaldarar, A. Prokofiev, E. Royanian, A. Amato, J. Sereni, W. Bramer-Escamilla, and I. Bonalde, *J. Phys. Soc. Jpn.* 76, 051009 (2007).

[3] E. Bauer and M. Sigrist (editors): *Lecture Notes in Physics* Vol. 847: Non-centrosymmetric Superconductors: Introduction and Overview, (Spring-Verlag, Berlin Heidelberg, 2012).

[4] N. Kimura, K. Ito, K. Saitoh, Y. Umeda, H. Aoki, and T. Terashima, *Phys. Rev. Lett.* 95, 247004 (2005).

[5] N. Kimura, Y. Muro, and H. Aoki, *J. Phys. Soc. Jpn.* 76, 051010 (2007).

[6] I. Sugitani, Y. Okuda, H. Shishido, T. Yamada, A. Thamizhavel, E. Yamamoto, T. D. Matsuda, Y. Haga, T. Takeuchi, R. Settai, and Y. Ōnuki, *J. Phys. Soc. Jpn.* 75, 043703 (2006).

[7] R. Settai, I. Sugitani, Y. Okuda, A. Thamizhavel, M. Nakashima, Y. Ōnuki, and H. Harima, *J. Magn. Magn. Mater.* 310, 844 (2007).

[8] G. Knebel, D. Aoki, G. Lapertot, B. Salce, J. Flouquet, T. Kawai, H. Muranaka, R. Settai, and Y. Ōnuki, *J. Phys. Soc. Jpn.* 78, 074714 (2009).

[9] F. Honda, I. Bonalde, K. Shimizu, S. Yoshiuchi, Y. Hirose, T. Nakamura, R. Settai, and Y. Ōnuki, *Phys. Rev. B* 81, 140507 (2010).

[10] S. S. Saxena and P. Monthoux, *Nature* 427, 799 (2004).

[11] V. M. Edel'stein, *Sov. Phys. JEPT* 68, 1244 (1989).

[12] L. P. Gor'kov and E. I. Rashba, *Phys. Rev. Lett.* 87, 037004 (2001).

[13] K. V. Samokhin, E. S. Zijlstra, and S. K. Bose, *Phys. Rev. B* 69, 094514 (2004).

[14] P. A. Frigeri, D. F. Agterberg, A. Koga, and M. Sigrist, *Phys. Rev. Lett.* 92, 097001 (2004).

[15] S. Fujimoto, *J. Phys. Soc. Jpn.* 76, 051008 (2007).

[16] V. K. Anand, A. D. Hillier, D. T. Adroja, A. M. Strydom, H. Michor, K. A. McEwen, and B. D. Rainford, *Phys. Rev. B* 83, 064522 (2011).

[17] V. K. Anand, D. Britz, A. Bhattacharyya, D. T. Adroja, A. D. Hillier, A. M. Strydom, W. Kockelmann, B. D. Rainford, and K. A. McEwen, *Phys. Rev. B* 90, 014513 (2014).

[18] E. Parthé, B. Chabot, H.F. Braun, and N. Engel, *Acta Crystallogr. B* 39, 588 (1983).

[19] A. J. Drew, S. L. Lee, F. Y. Ogrin, D. Charalambous, N. Bancroft, D. McK. Paul, T. Takabatake, and C. Baines *Physica B* 374-375, 270 (2006).

[20] W. McMillan, *Phys. Rev.* 167, 331 (1968).

[21] T. Terashima, M. Kimata, S. Uji, T. Sugawara, N. Kimura, H. Aoki, and H. Harima, *Phys. Rev. B* 78, 205107 (2008).

[22] C. Kittel, *Introduction to Solid State Physics*, 8th Ed. (Wiley, New York, 2005).

[23] M. Tinkham, *Introduction to Superconductivity*, 2nd Ed. (Dover, Mineola, NY, 1996).

[24] B. Mühlschlegel, *Z. Phys.* 155, 313 (1959).

[25] D. C. Johnston, *Supercond. Sci. Technol.* 26, 115011 (2013).

[26] J. Bardeen, L. N. Cooper, and J. R. Schrieffer, *Phys. Rev.* 108, 1175 (1957).

[27] H. Padamsee, J. E. Neighbor, and C. A. Shiffman, *J. Low Temp. Phys.* 12, 387 (1973).

[28] V. K. Anand, H. Kim, M. A. Tanatar, R. Prozorov, and D. C. Johnston, *Phys. Rev. B* 87, 224510 (2013).

[29] V. K. Anand, H. Kim, M. A. Tanatar, R. Prozorov, and D. C. Johnston, *J. Phys.: Condens. Matter* 26, 405702 (2014).

[30] P. G. de Gennes, *Superconductivity of Metals and Alloys* (Benjamin, New York, 1966)

[31] D. Saint-James and P. G. de Gennes, *Phys. Letters* 7, 306 (1963).

[32] M. Strongin, A. Paskin, D. G. Schweitzer, O. F. Kammerer, and P. P. Craig, *Phys. Rev. Lett.* 12, 442 (1964).

[33] M. Smidman, A. D. Hillier, D. T. Adroja, M. R. Lees, V. K. Anand, R. P. Singh, R. I. Smith, D. M. Paul, and G. Balakrishnan, *Phys. Rev. B* 89, 094509 (2014).

[34] N. Kimura, H. Ogi, K. Satoh, G. Oshaki, K. Saitoh, H. Iida, and H. Aoki, *JPS Conf. Proc.* 3, 015011 (2014).

[35] Y. Okuda, Y. Miyauchi, Y. Ida, Y. Takeda, C. Tonohiro, Y. Oduchi, T. Yamada, N. D. Dung, T. D. Matsuda, Y. Haga, T. Takeuchi, M. Hagiwara, K. Kindo, H. Harima, K. Sugiyama, R. Settai, and Y. Ōnuki, *J. Phys. Soc. Jpn.* 76, 044708 (2007).

In: Superconductivity
Editor: Paulette Grant

ISBN: 978-1-63483-907-5

Chapter 2

ABOUT THE MECHANISM OF ARISING OF THE SUPERCONDUCTIVITY, INDUCED BY WATER IN COMPOUNDS $SrFe_2As_2$ AND $FeTe_{0.8}S_{0.2}$

Ludmila Stepanovna Martseniuk*
Institute for Nuclear Researches NAS Ukraine, Kiev, Ukraine

Abstract

$SrFe_2As_2$ and $FeTe_{0.8}S_{0.2}$ compounds are layered ferriferous compounds, which are perspective for developing new superconducting materials on their basis.

For these compounds, the effect of water-induced superconductivity was discovered. This effect occurs due to the diffusion of environmental water vapors into the compound after exposing the virgin samples of compounds to an ambient atmosphere at room temperature. Up to now, there has been no satisfactory explanation of this effect, but its occurrence is generally associated with compression of the compound crystal lattice as the result of water vapor diffusion. The cause of such compression is also not determined.

The analysis of a possible processes taking place in the quasi two-dimensional layers of diffused water, was conducted in this work with using of the theory of J. Preparata, who from the position of quantum field theory, presented the description of the processes of water molecules condensation to liquid state in the case of usual water. It is shown that the properties of water in the quasi-two-dimensional structure are significantly different from the properties of usual water. The reasons on which such water can compress a crystalline lattice and induce the arising of superconductivity are also stated.

The properties of water diffusing into $SrFe_2As_2$ and $FeTe_{0.8}S_{0.2}$ compound are compared with the properties of nanostructure water in carbon nanotubes (water with a quasi-one-dimensional structure) at low temperatures.

* E-mail address: prolisok77@yandex.ua; Fax: (044)4125471; Address: Institute for Nuclear Researches NAS Ukraine, 03680, Prospect Nauky 47, Kiev, Ukraine.

1. Introduction

The advance in modern superconductivity physics is impossible without formation of clear conceptions about the character of the processes occurring in the superconducting materials at their transition to a state of superconductivity. The discovery of new perspective high-temperature superconducting materials has caused problems in the interpretation of properties of such materials, as these properties cannot be explained by using traditional methods. At the present time, there is no theory describing satisfactorily the processes in high-temperature superconducting materials. Actually, every new paper in connection with the study of properties of such materials proposes new ideas, which contradict to, or significantly modify the previous concepts. So, according to the existing theories, there is no consensus about the mechanism of processes in high-temperature superconducting materials.

This statement specifically applies to the effect, newly discovered by Japanese researches, of superconductivity in $SrFe_2As_2$ and $FeTe_{0.8}S_{0.2}$ compounds induced with water or an alcohol-water solution [1-3]. At the present time, there is no acceptable explanation for this effect. This fact testifies that the new non-conventional methods, which are being developed in the fields of high-temperature superconductivity theory and physics of water, are required for the explanation.

It should be noted that the practical and theoretical studies of the above-mentioned problems can result in new outlooks in studying nanowater, such as water in carbon nanotubes, water in capillary blood vessels, water in plant capillary systems, and near-surface water. Additionally, such studies can promote further theoretical works in the fields of physics of superconductivity in layered compounds and physics.

2. The Effect of Water-Induced Superconductivity in $SrFe_2As_2$ and $FeTe_{0.8}S_{0.2}$ Compounds

2.1. Structural Features and Properties of $SrFe_2As_2$ Compound. The Processes Occuring at Diffusion of Water in This Compound

The $SrFe_2As_2$ compound belongs to oxypnictides and, as all oxypnictides, can be superconducting only when doped with electron or hole impurities or at application of high pressure [1]. As the authors [1] discovered, there is another method for inducing superconductivity in this compound, which consists in exposing the compound to the open air for several hours in order to induce the diffusion of environmental water molecules into the compound. In their studies, the authors [1] used undoped epitaxially-grown films, about 200 nm in thickness, produced by using a laser evaporation method. The schematic representation of the crystal lattice of AFe_2As_2 compound (where A is Ba, Sr, or Ca) is shown in Figure 1 [4]. As is pointed in [1], the structure of $FeAs$ layers of $SrFe_2As_2$ compound is similar to the structure of other superconducting ferriferous compounds of this type.

The $FeAs$ layers consist of $FeAs_4-$ tetrahedrons, which are essential for forming the superconducting state in superconducting materials.

Author [1] pointed that undoped $SrFe_2As_2$ “exhibit structural phase transitions, which accompany antiferromagnetic ordering transitions at low temperatures but do not exhibit superconductivity at least down to 2 K.”

The results of additional studies performed by the authors [1] have demonstrated that only water diffused into the epitaxially-grown films of $SrFe_2As_2$ compound, which were held in the open air, induced the superconductivity in the studied compound.

The critical transition temperature T_c is 25 K, that is, is higher by 5 K than the critical temperature for the samples doped with cobalt. According to [1], this means that the process of transition to the superconducting state for the samples exposed in the open air with water vapors differs from the process for the samples doped with cobalt.

The authors [1] noted that water-intercalated superconductivity was detected in some other layered crystals such as:

$$A_x(H_2O)_yTS\ (A = Na, K, Rb, Cs;\ T = Nb, Ta;\ T_c = 5.5K);$$

$$Na_xCoO_2 \cdot yH_2O\ (T_c = 5K);$$

$$Na_x(H_2O)_y(MS)_{1+\delta}(TaS_2)_2\ (M = Sn, Pb, Sb;\ T_c = 4K).$$

At the diffusing of water into these compounds, the distance between the basic layers of the compound is increased and superconductivity is promoted, apparently, due to the enhanced two-dimensionality caused by the increase of the inter-layer distances as the result of the water diffusion.

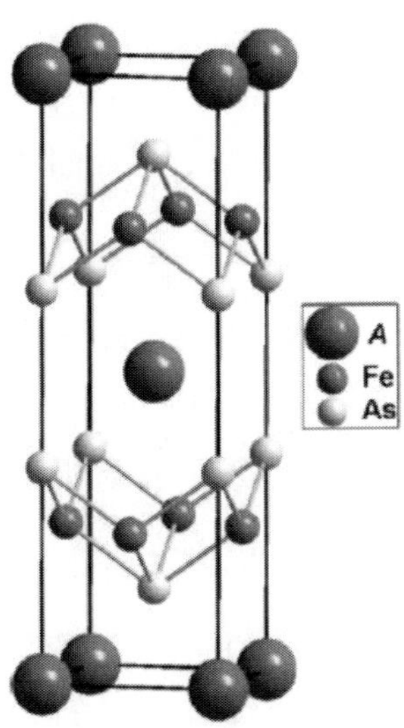

Figure 1. Crystal lattice of AFe_2As_2 (where A is Ba, Sr, or Ca) [4].

The opposite effect is discovered at the diffusion of waters into $SrFe_2As_2$ epitaxially-grown films. This effect is characterized by the compression of the compound lattice in parallel to C axis, as well as strong magnetic anisotropy, which significantly exceeds the magnetic anisotropy for the $SrFe_2As_2$ alloyed with cobalt.

The authors [1] state that: "the *c*-axis length of the $SrFe_2As_2$ phase was decreased from 1.233 nm in the virgin film to 1.227 nm in the air-exposed film. The latter value is smaller even than that of the cobalt-doped $SrFe_2As_2$ epitaxial films (c = 1.230 nm)."

Additionally, temperature T_c and superconductivity anisotropy are significantly different for the $SrFe_2As_2$ compound doped with aliovalent ion impurity [1].

When external magnetic field $H \approx 9\ T$ is applied, the superconducting transition temperature T_c is not higher than 7 K for H//c and not higher than 16 K for H//a. The authors [1] are of opinion that the upper critical magnetic fields are far above 9 T and the temperature T_c is more sensitive to the H|//c than to the H//a.

The superconductivity anisotropy for $SrFe_2As_2$ compound doped with cobalt is virtually not detected when the direction of magnetic field changes.

Another characteristic feature noted by the authors [1] consists in the presence of some quantity of Fe_2As impurity phase, which can be detected by using X-ray structure analysis procedure. This feature implies that water diffusion causes some reconstruction of an internal molecule structure in the basic layers of $SrFe_2As_2$ compound, but without increasing the distance between the basic crystal layers.

Considering that there are no data in scientific and technical literature about the transition of Fe_2As phase into the superconducting state, the authors [1] are of opinion that the formation of the superconducting transition is basically caused by compression of the crystal lattice, which is equivalent to external pressure action, rather than by the additional Fe_2As phase. This conclusion is based on the data that external pressure induces superconductivity in $SrFe_2As_2$ compound at temperature T_c = 35 … 38 K. But this method of inducing superconductivity is characteristic by the decrease of temperature of transition to antiferromagnetic state $T_{ant.}$ at the increasing of pressure.

For $SrFe_2As_2$ epitaxially-grown films containing diffusing water, temperature $T_{ant.}$ is the same as for the virgin samples, and temperature T_c is lower than that in the case of external pressure. It means, that the mechanism of occurrence of superconducting transition other, than at external compression.

Studies of superconductivity in $SrFe_2As_2$ compound under external pressure are described in some papers, including [4-6].

The studies performed by the authors [5] for undoped $SrFe_2As_2$ compound have demonstrated that compression of the crystal lattice under pressure causes a new effect which was earlier detected neither for $SrFe_2As_2$ compound alloyed with cobalt nor for any other iron-containing superconducting compound.

The authors [5] studied, by using a nuclear magnetic resonance analysis method, the monocrystal samples of $SrFe_2As_2$ compound exposed to an external pressure of up to 9.4 GPa. High pressure is suppressive for the antiferromagnetic state.

The authors [5] detected the simultaneous presence of two phases – superconductive (SC) and antiferromagnetic (AF) at external pressures in the range of 4.2 - 7.2 GPa.

The similar processes were not detected for $SrFe_2As_2$ compound exposed in the open air. Thus, the mechanism of transition to superconductivity of $SrFe_2As_2$ compound at diffusion of water molecules into crystal lattice cannot be connected only with the compression of a lattice.

2.2. The Features of the Effect of Inducing by Water and by Alcogolic Beverages of Superconductivity in $FeTe_{0.8}S_{0.2}$ Policristal

The effect of inducing by water of superconductivity in compound $SrFe_2As_2$, as well, as the compression of a crystal lattice, as appeared, are possible not only in $SrFe_2As_2$ epitaxially-grown film samples. The authors [2] for the first time discovered the analogous effect in $FeTe_{0.8}S_{0.2}$ polycrystals.

As the authors [1], the authors of [2] could not adequately explain the processes of inducing of superconductivity by water as well as the compression of the crystal lattice of $FeTe_{0.8}S_{0.2}$ polycrystals occurring in the result of absorbing water vapors from the ambient air at room temperature.

The schematic representation of the crystal lattice of the $FeSe$ compound close to $FeTe$ compound is shown in Figure 2 [7]. Both compounds have the same structure.

The authors [2] indicate that transition to the superconducting state under external pressure, with the onset temperature T_c^{onset} increasing from 13 K to 37 K at a pressure of 4 … 6 GPa, was detected in $FeSe$ compound with a tetragonal structure, but was not detected in undoped telluride of iron. At a temperature of 70 K, tetragonal telluride of iron $FeTe$ undergoes antiferromagnetic ordering and does not show superconductivity.

To suppress the magnetic ordering, the authors doped ferrous telluride with sulfur. As a result, $FeTe_{0.8}S_{0.2}$ polycrystals were produced.

As is noted [2], the properties of $FeTe_{0.8}S_{0.2}$ compound depend on methods of the crystal growing process. For their studies, the authors used samples produced according to the solid-state reaction process. The samples were produced close to the single-phase compound samples.

For the as-grown samples, zero resistivity was not observed while an onset of the superconducting transition was observed at 8.0 K. The proper to superconductivity diamagnetic signal was not found out.

However at exposing of the samples to air at a room temperature, the polycrystals have changed the characteristics and the transition to a superconducting state has been fixed for them. The temperature of transition to superconductivity depends on a time of exposing of samples to air, as shown on Figure 3.

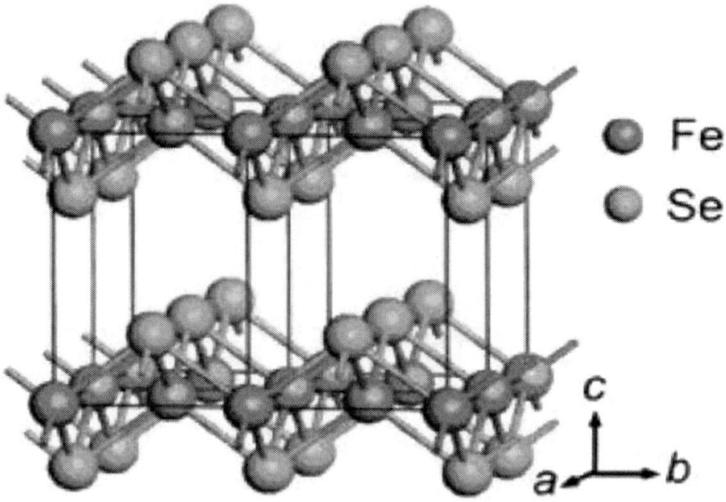

Figure 2. Crystal structure of *FeSe* - compound [7].

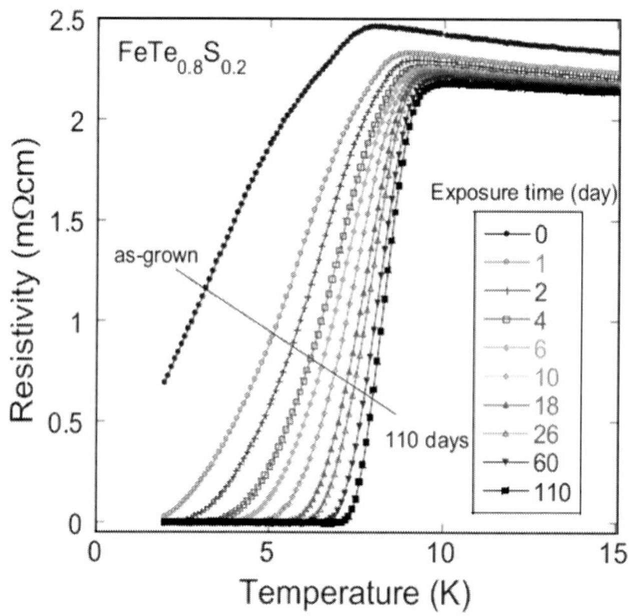

Figure 3. Temperature dependence of resistivity for the $FeTe_{0.8}S_{0.2}$ sample kept in the air for several days. The color of curves designates the time of an exposition to air [2].

The highest temperature of transition was recorded for the samples exposed to air during 110 days. After exposing the sample to the air for 2 days, zero resistivity appeared around T_c = 2K. With increasing air-exposure time, both T_c^{onset} and T_c increased up to 10.2 K and 7.2 K, respectively.

In addition, appeared, that “the normal-state resistivity of 110-day-old $FeTe_{0.8}S_{0.2}$ is clearly lower than that as grown, implying the change in the carrier density by the air exposure” [2]. The other change of polycrystal properties, which is caused by diffusion of water molecules, is the change of the polycrystalline lattice parameters [2]. It is reported in [2] that the virgin synthesized samples and the samples held in the open air for 100 days and 200 days were examined by using the X-ray structure analysis procedure.

For 200-day-old $FeTe_{0.8}S_{0.2}$: "Lattice constants were calculated to be a = 3.8158(9) and c = 6.2445(23) Å for as-grown $FeTe_{0.8}S_{0.2}$, a = 3.8114(8) and c = 6.2421(21) Å for 100-day-old $FeTe_{0.8}S_{0.2}$, and a = 3.8097(8) Å and c = 6.2307(20) Å for 200-day-old $FeTe_{0.8}S_{0.2}$, respectively." The authors [2] could not determine the cause of the water-induced superconductivity in the $FeTe_{0.8}S_{0.2}$ compound, but are of opinion that: "Moisture-induced superconductivity in $FeTe_{0.8}S_{0.2}$ might be induced as in $SrFe_2As_2$."

The authors [3] reported about the more interesting effect consisting in that the superconductivity in $FeTe_{0.8}S_{0.2}$ polycrystals can be induced with the alcohol-water solution even more efficiently. The researches of resistivity and magnetic properties have shown, that the zero resistivity temperature (T_c^{zero}) of the $FeTe_{0.8}S_{0.2}$ samples is weakly dependent on the type of alcoholic beverages. This temperature is in the range of 7.1 K for solution containing Japan sake to 7.8 K for solution containing red or white wine.

The volume of a superconducting state is more dependent on the type of alcohol-water solution. The authors determined that the most affective medium for superconductivity is red wine. In this case, the volume, occupied by superconductivity is increased by 10-60% as compared with superconductivity induced with usual water.

3. About of the Nature of Appearance of Superconductivity in Ferriferous Superconductors

As is known [8], the impurity of magnetic materials negatively influence upon parameters of a superconducting state of usual superconductors. It is stipulated by the fact that magnetic impurity possesses its own magnetic moment, which interacts with the magnetic moment of one of Cooper pairs electrons and reorients its direction, which destroys the coupled state of pair.

The existence of high-temperature superconducting ferriferous compounds was revealed in 2008 and, despite of huge number of researches, the mechanism of pairing of electrons in these compounds is not opened. The standard theory BCS, created for the description of processes in earlier known superconductive compounds, is not sufficient as it denies the possibility of origin of superconductivity in magnetic materials [8]. Nevertheless, as follows from references, exactly magnetism can be responsible for the unusual mechanism of superconductivity in these compounds.

As the majority of these compounds have phase transition in an antiferromagnetic state, there is an assumption that the antiferromagnetic state precedes superconducting, and they compete among themselves.

Usually to suppression of an antiferromagnetic state apply doping. Compression changes parameters of a crystalline lattice that influences magnetic ordering and also can suppress antiferromagnetism. The theoretical development of models of high-temperature superconductivity resulted in the appearance of a wide spectrum of various assumptions; however the theory of origin of superconductivity in these compounds is far from completion.

At first, in connection with discovering new superconductors, the possibility of using the standard theory of electron-phonon interactions for explanation the mechanism of Cooper pair's formation was researched. So, in work [8] it has been shown theoretically, that in using the model of electron-phonon interactions, the temperature of transition in a superconducting state for iron chalcogenides, to which pertains and $FeTe$, will not exceed 1 K. Taking into account, that the value of critical temperature is much higher, it means, that the given compounds belong to superconductors with the unusual mechanism of pairing. In this work it is also shown, that properties of compounds $FeSe$, $FeTe$ and FeS are similar by configuration of a Fermi's surface, phonon spectrum, and magnetism with compounds on a basis $FeAs$. Authors consider that superconductivity is caused by spin fluctuations of spin density waves (SDW) type. In work [8] calculated data about the structure of the Fermi surface are discussed. It is shown, that Fermi's surface has a complicated character and consists from cylindrical electronic section in corners of Brillouin zone and a section of cylindrical hole surfaces in the center of a zone. Such structure is caused by a quasi-two-dimensional electronic structure of ferriferous superconductors.

The further experimental researches, which have been carried out by various methods, have confirmed these calculations.

For estimation the possibility of electron-phonon mechanism of coupling, the researches of the isotopic effect in these compounds [9] has been carried out. If takes place the electron-phonon mechanism, dependence of the critical temperature on mass of an isotope (at replacement of a superconductor isotope) should be inversely proportional to a square root of an isotope mass m, i.e., $T_c \sim 1/m^{\alpha}$ where $\alpha = 0.5$.

For an estimation of parameter α one usually uses a more convenient formula:

$$\alpha = -\frac{m}{\Delta m}\frac{\Delta T_c}{T_c} \tag{1}$$

For the compound $Ba_{1-x}K_xFe_2As_2$ (x = 0,4) with $T_c = 37\,K$ the researches of isotopic effect have been carried out and it appears, that at replacement of iron ^{56}Fe by an isotope ^{54}Fe the value α is approximately equal to 0.35. In the opinion of authors, it means that the electron-phonon mechanism of the formation of Cooper pairs is possible, but it is not described by the standard Barden, Cooper and Schrieffer (BCS) theory.

One more work [10] also devoted to research of isotopic effect, has not confirmed results of work [9]. They researched the same polycrystalline compound $Ba_{1-x}K_xFe_2As_2$, and at the isotopic replacement of atoms of iron, it was revealed that though the exponent α differs from zero, it has a negative sign and consists of $\alpha_{Fe} = -0.18$. It means that the heavier an isotope of iron, the higher value of critical temperature.

Based on the above authors [10], proceeding from received result, believe, that electron-phonon interaction participates in formation of superconductivity, but character of a display of such interaction is complex and is not described by standard BCS model. Possible also vary, when the mechanism of formation of Cooper pairs is represents a mix of exchange and electron-phonon interactions.

Interesting in this respect is the already mentioned work [9]. In this work authors investigated not only the influence of isotopic replacements on T_c, but also investigated the changes of the temperature of transition of not doped compounds $BaFe_2As_2$ and $SmFeAsO$ in a state of spin density wave T_{SDW}. It appeared that $\alpha_{SDW} = -\frac{d\ln T_{SDW}}{d\ln m}$ is 0.05 for oxygen and 0.37 for iron. These data directly specify the presence of strong phonon-magnon interactions. In connection with these results it is necessary to suppose, that an isotopic change of value T_c is also caused by phonon-magnon interaction.

An important point for the opening of the mechanism of superconductivity is the establishing: which type of electrons behaviour in superconductor compound are realized, - the mechanism of strong binding according to which electrons can be distinguished, or the mechanism of weak binding, when electrons behave as united integer (so-called strongly correlated the Mott's dielectric).

As follows from review's work [11], the most popular standpoint, that $FeAs$ - systems are weakly correlated systems. However there are also a number of experimental works whose results do not agree with such a statement.

The author [11] in this connection indicates that though a question on whether $FeAs$ are weakly correlated systems or are strong correlated systems is remains open, electronic correlations play a certain role in these systems. The investigation of properties of iron chalcogenides (FeX) have shown, that, as well as for compounds such type as AFe_2As_2 (where $A = Ca, Sr, Ba$), to transition of chalcogenides to a superconducting state is precedes the phase transition to an antiferromagnetic state.

Suppression of an antiferromagnetic state is possible to achieve by doping or compression due to application of the external pressure.

Antiferromagnetic ordering in undoped ferriferous superconductors exists as waves of spin density (SDV), which it is possible, in the simplified approach, to associate with periodic reduction and increasing of electronics spin density. The period of these waves is determined not by the period of a lattice but by properties of magnetic interaction and this state is distinct from variety of magnetic excitation - spin waves.

As SDW-waves arise as result of magnetism in the system, which is caused by exchange interaction, the mechanism of formation of Cooper pairs in these compounds also can have an exchange character. In the works, where investigated the dynamics of spin excitation [12, 13] on not doped $SrFe_2As_2$ and $BaFe_2As_2$ by methods of neutron diffraction analysis, it is informed, that at low temperatures antiferromagnetic ordering takes place. For $SrFe_2As_2$ [12], it is revealed, that the spectrum of magnetic excitations has a gap $\Delta \leq 6,5$ meV, above which spin waves are observed. Magnone spectrum spread up to energy of order 170 meV, essentially exceeding typical frequencies of phonons (40 meV) [13].

Despite of huge quantity of publications on revealing the mechanism of pairing in high-temperature superconductors, there are some unsolved questions. As it was noted in the review [11], there are some unclear questions: "Is 'vicinity' to antiferromagnetism needed for high-temperature superconductivity? Whether antiferromagnetism is a competing phase, or does it substantially promote high-temperature superconductivity, for instance, through replacement of electron-phonon mechanism of pairing by the mechanism, based on spin (antiferromagnetic) fluctuations"? If the author [11] consider that in ferriferous superconductors "a role of electron-phonon interactions, apparently, can be quite significant (isotope - effect)," the author of another review [7] asserts "the vicinity to antiferromagnetic ordering in cuprates and *FeAs* - systems stipulates the spin-fluctuating mechanism of the pairing resulting in both cases to high values T_c."

We shall note that for high-temperature superconductors the mechanism of transition to superconducting state can appear even more complex than it was supposed and have a mixed character.

This point of view is also confirmed by work [5], where authors investigated influence of pressure on parameters of transition in a superconducting state of $SrFe_2As_2$ monocrystal. This work is interesting and in that relation, that induction by water of superconductivity in $SrFe_2As_2$, as mentioned above, explained by compression of a lattice, which occurs and at application of the external pressure.

In this work experimentally, by a method of a nuclear magnetic resonance for the first time the simultaneous existence of two phases - antiferromagnetic and superconducting (Figure 4) - for undoped $SrFe_2As_2$ with the application of external pressure in a diapason 4,2 - 8 GПa has been found.

As can be seen from Figure 4, in the range of 4.2-8 GPa the curves of Neel and phase transition in superconducting state coincide. It means that in this area there is a possible simultaneous existence of superconducting and antiferromagnetic phase.

Such data indicate, that antiferromagnetic state and superconducting state cannot only compete one with others, but also coexist.

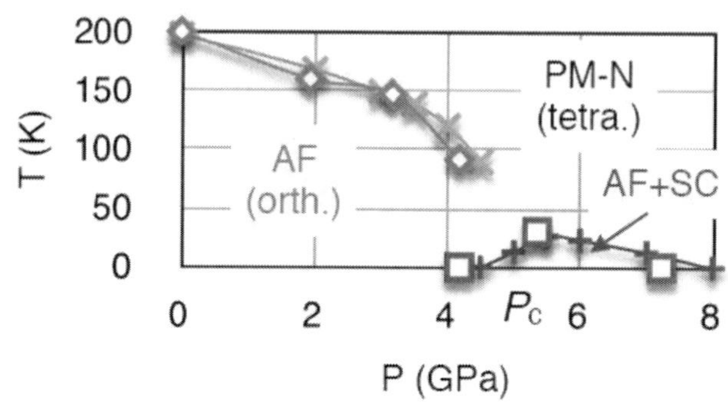

Taken from [5].

Figure 4. The phase diagram of $SrFe_2As_2$ compound depending on pressure. *Upper curve* divides areas of antiferromagnetic and paramagnetic states; the bottom curve outlines area of simultaneous existence of antiferromagnetic and superconducting states.

Besides, these data point to significant role of magnetic interaction (having character of exchange interaction) in realization of the mechanism of superconductivity in this compound.

Authors also note that the effect, observable by them appears in consequence of formation of nano-dimensional superconducting/antyferromagnetic hybrid structures and this entangled state is a remarkable example of self-organizing of heterogeneous structures in pure systems.

Let us note also that this work [5] to some degree approximates us and to comprehension of essence of the mechanism induction by water of superconductivity in films $SrFe_2As_2$.

Thus the mechanism of arising of a superconducting state in ferriferous superconductors still not is opened; the attempts of using of any theory, suggested for the description of properties ferriferous superconductors, to explain the reason of stimulation by water of superconductivity in $SrFe_2As_2$ and $FeTe_{0.8}S_{0.2}$, were not successful also.

The problems which arose up at the searches of mechanism of effect of the inducing by water of superconductivity in compounds $SrFe_2As_2$ and $FeTe_{0.8}S_{0.2}$, can be caused by that the researchers in own interpretations did not aspire to take into account the properties of water, which as a result of diffusion in these compounds, most likely, is located by the single layers in parallel to the layers of basic material.

The properties of such quasi-two-dimensional water are significantly different from the properties of usual volumetric water, as is demonstrated in various papers concerning studies of quasi-two-dimensional water, nanostructure water, water in carbon nanotubes, water in molecular systems of living organisms, and near-surface water [14].

4. Some Positions of the Theory of Water Molecules Condensation in the Liquid State, Formulated in [9] on the Basis of a Quantum Field Theory

Water, in spite of its simple formula, is a very complex substance and behaves differently from other liquids, at diffusion of alien substances, at the dependence of density from temperature, and because of its many properties, including the nature of all its electric constants.

The traditional theories which were usually used at the description of water properties were started from assumption that water has homogeneous structure and exist as a single-phase system. However it was found out, that water revealed a variety of properties which cannot be explained by traditional theories, it testifies to insufficiency of such theories.

The theory of J. Preparata [15], created for the description of the phenomena in volumetric water, has enabled to explain well those phenomena which in principle could not be the described by the traditional theories supposing of a homogeneous structure of usual water, as these phenomena directly testified about the insufficiency of such a single-component model.

Let's stop on some positions of this theory.

In accordance with [16, 17], the state of the quantum system is determined as a coherent, if it is characterized by the physical variable named the phase ϕ of the field.

The quantum field is characterized by amplitude Ψ_0 in which square is proportional to number of quants, and by a phase ϕ:

$$\Psi = \Psi_0 e^{i\phi} = a\sqrt{N}e^{i\phi}, \quad (2)$$

In case of water system, N is a number of atoms in the selected volume.

Simplifying the description of forming process of coherence regions from the position of quantum field theory, it is possible to do as follows [17]:

We will suppose that there is a certain ensemble from N molecules having two states - the ground states and the excited states with the energy of excitation $E=\hbar\nu$ being approximately 12 эВ. The size of a molecule about 0.1 nanometers, and the size of a virtual photon with length of a wave $\lambda=c/\nu$, which can make excitation with energy 12 эВ, should be about 100.0 nanometers.

Therefore the linear size of region on which the excitations can be in 1,000 times more than a molecule. Let's assume that one virtual photon which is fluctuated from vacuum can excite an atom with probability P (according to a rating taken from the data at Lembov's shift value P about $10^{-4} \div 10^{-5}$). The excited molecule will be recombined through a certain time of a recombination, returning back the photon in an environment which will be either absorb by vacuum or will excite other molecule.

The corresponding probability of these two events will depend on the density $n=N/\lambda^3$ of the molecules, present inside of volume λ^3. When the density n will exceed threshold value n_{crit}, the photon can never return to vacuum and will pass from one molecule to another inside volume λ^3 under the condition of:

$$P\lambda^3 n_{crit}=1. \tag{3}$$

Thus, a vacuum gives a photon in a matter. This process will proceed until the great number of photons will not be captured and enough substantial electromagnetic field in this region does not appear.

The consequence of the formation of this field will be the occurrence of an attraction between molecules, to which short-range force of intermolecular repulsion will counteract, and the formation of the common oscillation of all captured molecules inside all regions; for this reason, the specified region is named as coherence domain (CD). In this process the molecules and photons lose the initial identity. Photons (as shown in these works) in the limited CD space acquire the imaginary mass; therefore cannot return back in a vacuum.

The regions designated as CD are surrounded with the molecules of "not coherent water."

At a common temperature, in accordance with the estimations, made in [16, 17], the ratio describing a requirement of coexistence of both fractions, as in liquid helium, is carried out:

$$F_c+F_{nc}=1 \tag{4}$$

The equation (4) determines for every value of temperature the general amount of the molecules, belongings to each phase; thus, each molecule passes through two phases continuously. It generates a "flickering" regime.

The temperature collisions, as well as for liquid helium, are the reason of existence of such regime. At room temperature, the total volume of such CD makes about 40% from total volume of water. At temperature near zero, the total volume of CD makes the most part of the volume of water and at approximately 100°C aspires to zero. Within the regions of CD, the property of water sharply differs from properties of noncoherent water: the viscosity of water medium and the decrement of vibrations are sharply reduced, and the fluidity of liquid is substantially increased. The situation will change for near-surface water (so-called EZ-water) and also in alive organisms where water closely contacts to surfaces of alive organisms (bio-membranes, bio-molecules). Appropriately to this case there will be a stabilization of the coherent state by virtue of attraction of water molecules to the wall, and a power barrier from the dividing effect of temperature collisions appears. In this case the coherent states can exist for a long time, and a formation of power whirlwinds in CD is essential. Just the same phenomena G. Pollak observed for, marking that for its formation, some forces of long-range action are responsible, and those molecules in it are in the excited state [18]. Let's note that a viscosity of near-surface water is essentially more than for usual water, and for the water bounded by molecular structures of an alive organism, viscosity, on the contrary, is reduced. In the effect explained in [17], in living organisms, the coherence is set not only within the regions of coherent domain but also between domains.

5. The Properties of One-Dimensional and Two-Dimentional Nanowater

For some system consisting of the elements interacting among themselves, the correlation relations between elements in the *classic* approach it is possible to assimilate with some springs which link the separate elements of the system. The action on one of the elements will cause the change of the state of other. However such changes do not occur instantly.

Let's consider from a position of traditional performances the conductivity of water which, as is known, connects with the proton jump conduction [19].

The scheme of the arrangement of the water molecules linked among themselves in a chain through a system of a hydrogen binding, is shown in Figure 5 taken from [19].

The proton in such chain of water molecules can be in one of the two equivalent positions (states) corresponding to position of a proton in that or other quantum well, shared by a quantum barrier.

For overcoming this barrier proton should receive the additional energy equal or large of the heights of a potential barrier, i.e., to pass in the excited state. However it can pass in another position and due to tunneling. In common case of the random distributing of the potential, the transparency of barrier is determined from the equation [20]:

$$D = D_0 \exp\left(-\frac{2}{\hbar} \int_{x_1}^{x_2} \sqrt{2\mu(U(x) - E)} dx \right) \tag{5}$$

where E is the energy of proton, μ is the mass of proton.

The proton owing to tunneling periodically passes from one position to another, i.e., it periodically changes the position. At application of a potential, proton jump conduction for volumetric water arises. This situation takes place for volumetric water. However for near-surface waters of alive organisms, for water forming the single nano-layers or disposing in nano-systems, nanotubes, and capillaries, and in some other cases, additional effects related to quantum entanglement can arise.

In connection with these problems, it is necessary to answer the following questions:

1. To what extent the traditional theories for describing the properties of water in quasi-two-dimensional systems formed by the diffusing water molecules in $SrFe_2As_2$ and $FeTe_{0.8}S_{0.2}$ compounds, can be used?
2. What mechanism of transferring of charge carriers in such systems at low temperatures?

It is possible to suppose, that there can be the additional effects connected with a quantum entangling, by analogy with the processes in such water systems: the separate water layers or the water chains in nanosystems, in nanotubes, or capillaries.

As an example of the entanglement, authors [21] have taken the system consisting of two particles with a one-half spin, which can accept only one of the two directions: "upwards" or "downwards." Such system is in a state with full spin, equal to zero (EPR state). This state is described by the formula:

$$|\Psi_{EPR}\rangle = \frac{1}{\sqrt{2}}\left(|\uparrow\rangle_1 \otimes |\downarrow\rangle_2 - |\downarrow\rangle_1 \otimes |\uparrow\rangle_2\right) \quad (6)$$

where: $|\uparrow\rangle_1, |\downarrow\rangle_2$ - wave functions of particles, one of which has a spin, directed upwards, and other the spin, directed downward. The symbol $\otimes$ designates the operation of direct (tensor) product. At the change of spin direction of the first particle the second particle also simultaneously must veer the spin direction, because the complete spin of the system is equal to zero.

The equation (6) describes the entangled state of two particles, but it is possible to describe the entanglement by analogical equation in the system consisting of a great number of particles.

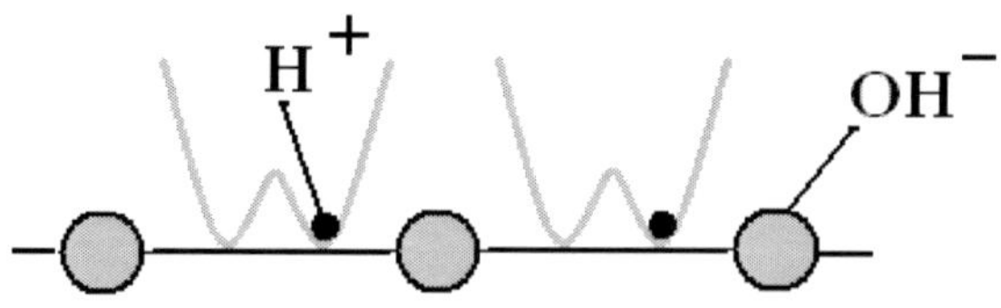

Figure 5. The scheme of the water molecules linked among themselves by the hydrogen bindings [19].

It is possible that exactly the quantum entanglement between the separate layers of water in compounds $SrFe_2As_2$ and $FeTe_{0.8}S_{0.2}$ is one of the factors of stimulation of superconductivity. In the case of entanglement between the molecules of water, linked in chains via the hydrogen binding, the change of a state of one molecule, will instantly lead to change of a state of other molecules.

If protons in any group of molecules appear in the excited state, sufficient for overcoming a quantum barrier sharing them, there will be the effects determined by not classical, but quantum laws. As specified in [22, 23] energy, necessary in order that a proton passed to the delocalized state, only on 20% higher than energy of hydrogen binding.

The effect for water at low temperatures in carbon nanotubes [24, 25], discovered by A. Kolesnikov with employees, has confirmed an opportunity of such delocalized state. Water in nanotubes is not linked with the surface of nanotube by virtue of its hydrophobic state.

In the works of these researchers it is specified that "The high momentum tail of the distribution, characteristic of the molecular covalent bond, is not present. ... Our data reveal that the protons in the hydrogen bonds are *coherently* delocalized and that the low temperature phase is a qualitatively new phase of ice." "New water" did not freeze even at a temperature, only on eight degrees distinguished from absolute zero.

The important assumption about the properties of nanotube water was made by the author in [22, 23]. The author proceeded from that fact that in a coherent state the electrons of water molecules are in the excited state and, consequently, protons also should be excited. Quantum barriers do not hinder the motion of such protons (Figure 5), and, in result, proton superconductivity is possible. The author [22, 23] refers to the works of [26, 27] and informs that in these works the behaviour of water in nanotube by diameter < 5 nm was investigated. Authors [26, 27] discovered that the property of such water "completely different from the more ordinary looking water confined in larger nanotubes." On the basis of experimental results of these works, author [22, 28] asserts: "I suggested that the water confined in the small nanotubes, being far more ordered, could be superconducting because jump conduction could occur simultaneously down multiple chains of hydrogen-bonded water molecules" [28]. The statement that water in a near-surface layer is in the excited state, was put forward also and by G.H. Pollak [18], therefore it is possible that it will be fair and for water, located by single layers in compounds $SrFe_2As_2$ and $FeTe_{0.8}S_{0.2}$.

6. About the Reasons of the Inducing by Water of Superconductivity in Compounds $SrFe_2As_2$ and $FeTe_{0.8}S_{0.2}$

As follows from [1, 2], at diffusion of water the compounds $SrFe_2As_2$ and $FeTe_{0.8}S_{0.2}$ essentially change properties and the most substantial change is the transition to the state of superconductivity at low temperatures, which is not found out for just made samples. Besides, as it was indicated above, there are also the structural changes at diffusion of water, one of which is compression of a lattice. Strong anisotropy of properties $SrFe_2As_2$ is found out at application a magnetic field at low temperatures, which arises for the sample, exposited in air, but does not reveal for doped of cobalt compounds [4]. The reduction of resistance [2] for the samples of $FeTe_{0.8}S_{0.2}$, exposed on air for more than 20 days, testifies to the increase in carrier's density at water diffusion.

A satisfactory explanation of processes, occurring in these compounds it is not found till now; researchers suppose that main reason of passing to the superconducting state is a compression of lattice, which is equivalent to the application of pressure [1, 3]. But for initial $FeTe_{0.8}S_{0.2}$, the structural transition to a superconducting state is not found out and at application the external pressure, and properties $SrFe_2As_2$ under pressure differ from properties of this compounds at diffusion in their of water.

Thus, any explanation put forward, constructed from comparison with properties of other high-temperature ferriferous superconductors, is not satisfactory. One other possibility remains, to explain the effect of induction on the basis of properties of the water, which are located, as result of diffusion, by single layers, paralleled to basic ferriferous layers.

Unfortunately, it is not known exactly, which one structures are formed by water at diffusing in compounds $SrFe_2As_2$ and $FeTe_{0.8}S_{0.2}$ of water molecules, but, nevertheless, the behaviour of this two dimension water is possible to predict, having applied to the description of properties of diffusing waters the positions, advanced by J. Preparata in the theory of condensation of water molecules to a liquid state [15], created by him on the basis of quantum field theory.

In [29] authors have theoretically shown the possibility of leaking out of photons through closely packed ensemble of atoms or molecules. As it is indicated in this work, it brings to continuous outflow of photons through mirrors of a cavity, even in the absence of an external influence. So, according to the theoretical estimation, resulted in this work, for 104 atoms ^{85}Ru placed in an optical cavity, emission of photons I = 300 c^{-1} is expected and this settlement value correspond to experimental data.

Authors [29] also indicated on possible processes of self-organization and the processes of outflow of energy, related to them, which can be caused by the interaction of the system of atoms (molecules) with virtual photons as it occurs in Cassimere effect. Authors also pay attention to a possible version of such system behaviour at presence of a dividing barrier and participation of the virtual photons coming from vacuum.

W. Nernst has shown in 1914 that the virtual photon can tune the vibration of identical molecules of the system and that provide arising of common phase of such vibrations in this system. This position is possible to use and for description of the behaviour of water molecules, diffusing in crystalline structure of compounds $SrFe_2As_2$ and $FeTe_{0.8}S_{0.2}$. Through the diffused into compounds of water molecules (in accordance with [29]) the virtual photons (from a vacuum) can leak also.

At existence of a sufficient transparency of the basic ferriferous layers of compounds in this energy diapason, the many water layers may be situated in region of interaction with a virtual photon. For this, the probability P of excitation (at a high enough degree of transparency) molecules of water layers, according to the equation (3), should be not less, than:

$$P_f = 1/\lambda^3 n_w, \tag{7}$$

where P_f - the value of probability of excitation by a photon a water molecule in the volume defined by size of a photon (~1000 Å);

n_w - the average density of molecules of water in this volume.

If ferriferous layer of the basic material (by virtue of some internal processes) greatly absorbs the radiation on frequency of such virtual photon, then in interaction is included the smaller number of the water layers or it occurs in single water layer only.

In the latter case it does not lead to those effects which, under our assumption, and are found out in $SrFe_2As_2$ and $FeTe_{0.8}S_{0.2}$ at diffusion in them waters (partly already at a room temperature, - for example, the compression of crystalline lattice, the increasing of the concentration of carriers).

The interaction of water molecules with the virtual photon, coming out from vacuum, can lead to the synchronization of vibrations of water molecules and formation of the coherence regions, similar to CD. According to positions of quantum field theory it will lead to an attraction between a molecules of water and, accordingly, between the layers of the molecules which have appeared in the field of interaction. The result will be the compression of a crystal lattice of compound.

Any other mechanisms of a lattice compression of compounds $SrFe_2As_2$ and $FeTe_{0.8}S_{0.2}$ at diffusion of waters in them, except the ones specified by us here (connected with the tune of vibrations of water molecules in a uniform by vacuum virtual photon), it has not been put forward till now.

It is interesting to indicate one more publication [3] where it is informed, that the effect of induction of superconductivity is found out and at contact of compound $FeTe_{0.8}S_{0.2}$ with the alcohol-containing water solution. While heating the $FeTe_{0.8}S_{0.2}$ compound in the various alcoholic beverages, with compound there are the same changes, as well as at an exposition on air, in particular, the compression of lattice. We shall note in this connection, that according to [30] the resonant spectra of EHF-diapason, which have been taken off from water and alcohol-containing water solution, have a very close location of a resonant peaks though the intensity of these peaks are different. Such distinction in [30] explained that for water, containing spirit, typically the delaminating of hexagonal planes was higher, than for pure water.

Ethanol spirit, as well as usual water, is an associated liquid, i.e., contains the hydrogen bindings. It means [3] that the positions of quantum field theory, advanced by G. Preparata for water [15, 17], are suitable for the description of this substance. Probably, exactly these characteristic also cause the possibility of molecules of an ethanol spirit to interact with virtual photon and induce the superconductivity in compounds $SrFe_2As_2$ and $FeTe_{0.8}S_{0.2}$ either as well as water molecules.

Though the unique reason of induction, in opinion of authors of the works [1, 2], which for the first time have fixed the effect, is the compression of a lattice (that equivalently as they consider, to application of external pressure), such explanation, as authors specify, is not sufficient.

In fact, the influence of diffusing waters is not reduced to compression only.

As follows from above-mentioned works, it is possible to induce the transition to superconductivity in two ways: first, compression and, second, doping [1, 2]. Doping can bring to the next result: to reduce the interaction between layers, to suppress the transition in an antiferromagnetic state and to increase the concentration of carrier.

Let us consider in more detail what occurs with the molecules of water, diffusing in compounds $SrFe_2As_2$ and $FeTe_{0.8}S_{0.2}$.

Water molecules, diffusing in these compounds, in consequence of interaction with a virtual photon and the subsequent passing to coherence, get such quantum characteristic, as a phase, unique for each region of coherence. The further interaction between separate regions can lead to their merging, as well as for EZ-water, especially at the decrease in temperature.

Let us note, that in the region of interaction with the virtual photon, having the energy $E = 10\ eB$ (which length of a wave, accordingly, makes about 1,000 Å), gets enough amount of layers of the water, which are settling down between the basic ferriferous layers. Thus, the condensation of diffusing water molecules in a state, on its properties reminding (at a low temperature) "ice-like" state, is carried out.

Let us emphasize here that the coherent water in a formations of water layers, as well as usual water in CD, is not the water, which properties correspond to frozen crystallize water, where coherent domains where absent. Here, we have a superposition of the ground state, where electrons closely bond among themselves and the excited, where electrons are "handling" in an electronic cloud and are collectivized for all region of coherence, i.e., are a unit. The most essential difference of water in two-dimensional layers from "ice-like state" is that: these layers do not couple in a transversal direction by hydrogen banding and have two "external" electrons on one atom of oxygen, because oxygen in "ice-like state" is tetravalent [30].

The second difference will be that in the field of interaction with a virtual photon for all water layers the regime of coherence is established, i.e., all processes occur in a uniform and with unique phase. Because in state of coherence between the layers of water there is the attraction, the parameters of crystal lattice can change. It is also observed experimentally.

Let us note, that at the usual freezing of water there is, as is known, a process of expansion, instead of compression (as in the case of the effect of induction by water of superconductivity in $SrFe_2As_2$ and $FeTe_{0.8}S_{0.2}$ compounds).

The third difference is that the polycrystal lattice stabilizes processes in water layers in accordance with their own internal process and can appear the resonance transitions (and it is also possible the processes, similar to some extent to effect of affinity, when superconductivity in one substance induces superconductivity in the other substance).

Diffusing water differs from EZ-water, as well as from usual water, by two-dimensional character, the absence of hydrogen binding between layers. At the same time inherent to it, as well as for EZ-water, higher degree of coherence, ability easily to give of electrons and, probably, number of other properties.

From the enumerated particularities of water in single layers follows that such layer:

1. Can shield interaction between layers of the basic material (it is expressed in occurrence of anisotropy of properties, which it was mentioned above for $SrFe_2As_2$),
2. Can be suppliers of electrons to the basic ferriferous layers (it is found out for $FeTe_{0.8}S_{0.2}$),
3. Can be factors of approaching of ferriferous layers (change of geometrical parameters [1, 2]), and
4. Can influence on the processes of ordering, than to suppress antiferromagnetic ordering, as compete with a superconducting phase (transition to superconductivity).

All these four listed factors are the factors promoting the transition in a superconducting state. However there is also one more factor, related to that at low temperatures the water can be only in the pair-state. For this state, as against from ortho-state, molecules of water cannot rotate and, consequently, to form the binding structures, the type of hexagonal [21].

As for such structures are available of two "external" electrons [30], the opportunity is not excluded, that these electrons also can participate in formation of Cooper pairs: or with others electrons of water layers, or with electrons from the basic ferriferous layers (taking into account, that the processes in individual water layers occur synchronously). We shall note that interaction of water molecules with atoms of iron can lead to conversion of water protons and transition of water molecules from a pair–state in an ortho-state [31] and thereby to promote the occurrence of processes of electrons pairing.

The synchronization of vibration of the water structures (for usual water is hexagonal structures) and of magnon processes of the basic ferriferous layers are possible also. Really, in [9] the presence of strong phonon-magnon interactions for compound $BaFe_2As_2$, similar to compound $SrFe_2As_2$, is revealed, and authors [30] find out the resonant vibrations of hexagonal water structures in a gigahertz range. Frequency of vibrations of a coherence water structures, located by a single layers, can be synchronized at low temperatures with the processes, caused by exchange interaction in iron-containing layers.

As there are no reliable literary data on finding of correlation interactions of superconducting type for electrons in layers of near-surface waters, (some indications on such opportunity for the water, bound by living structures, are available in [32]), the most probable mechanism of stimulation, besides possible processes of pairing electrons from layers of water and the basic ferriferous layers, apparently, will be a complex manifestation of the factors listed in points 1-4.

At decreasing of temperature, the volumes of coherency regions in quasi two-dimensional layers of diffused water will be increased. At very low temperatures, water layers can be in the state with properties which, as in the case with one-dimensional water chains, are different as compared with the properties of ice.

As in the case of near-surface water, protons in quasi-two-dimensional layers can be delocalized. This means that and in quasi-two-dimensional layers can be the conditions which, according to [22], promote the transition of water in nanotubes to the state of proton superconductivity. If this assumption is correct, the water in water layers of $SrFe_2As_2$ and $FeTe_{0.8}S_{0.2}$ compounds can be in the state of proton superconductivity. It is probably that such transition can coincide with the transition of $SrFe_2As_2$ and $FeTe_{0.8}S_{0.2}$ compounds into the superconducting state.

It is necessary to confirm this assumption by results of experimental studies performed in accordance with procedures developed for water in nanotubes.

Conclusion

At the present time, the effect of superconductivity induced by the water or the alcohol-water solution in $SrFe_2As_2$ and $FeTe_{0.8}S_{0.2}$ compounds has not been explained notwithstanding the thorough study of changes caused in these compounds by water diffusion. The authors [1, 2], which for the first time discovered this effect, stated that the diffusing of water causes compression of the compound crystal lattice, change of electric and magnetic properties, and the increasing of anisotropy of some parameters and their dependence on temperature.

The authors [1, 2] suppose that the most probable cause of water-induced superconductivity in $SrFe_2As_2$ and $FeTe_{0.8}S_{0.2}$ compounds is the compression of the compound crystal lattice, and that the compression is equivalent to external pressure action. The authors showed that there is significant difference between the properties of compounds exposed to external pressure and the properties of compounds held in the open air for the period of time sufficient for diffusion of water into the compound. Such difference indicates that there are different mechanisms for promoting superconductivity in these two cases.

The effect of compression of the crystal lattice with water is also not explained. It should be noted that the compression can be one of the basic factors promoting superconductivity in these compounds. The attempts of attracting of various theories to explain induced superconductivity in ferriferous compounds were ineffective.

So, when studying the effect of superconductivity inducing by water in $SrFe_2As_2$ and $FeTe_{0.8}S_{0.2}$ compounds, it is necessary to take into account various processes occurring due to diffusion of water in these compounds.

It is most probably that diffusing water is present in single layers arranged in parallel to the basic layers of $SrFe_2As_2$ and $FeTe_{0.8}S_{0.2}$ compounds.

As shown in the experimental works, which have been conducted by A. Kolesnikov with employees [24, 25], water in carbon nanotubes exists in the special state, and, actually, does not freeze up to 8 K. According to [24, 25] it is expected that water in quasi-two-dimensional layers (two-dimensional water) is also different from usual water.

For the description of properties of such water it is also possible to apply the theory of J. Preparata [15]. This theory was developed for description of water molecules condensation to a liquid state because of the interaction of virtual photon with water molecules. As a result the water medium contains the regions of coherence, - coherence domains (CD).

According to [29], a virtual photon can penetrate with the certain probability through the densely packed ensemble of atoms. If the region of interaction with the virtual photon contains a great number of water layers, these layers can form the regions of coherence, with the common phase for the each region of coherence.

This process, according to quantum field theory, induces the attracting between the molecules of water and between the layers of water (the analogous process is in the case of condensation of molecules) and the compression of the crystal lattice of $SrFe_2As_2$ and $FeTe_{0.8}S_{0.2}$ compounds as the result of water diffusion.

Other explanations of the crystal lattice compression have not been proposed.

According to this theory, water coherence regions are formed at room temperature and are increasing in volume when temperature decreases.

In such coherence regions, the following processes can occur:

1. The geometric characteristics of the compound crystal lattice can change as in the case of changes caused by external pressure [1, 2].
2. The interaction between the basic layers of the compound can be restricted, resulting in the anisotropy of properties as is mentioned above for $SrFe_2As_2$ compound [1].
3. The water layers can be affective sources of electrons for the basic ferriferous layers, as is detected for $SrFe_2As_2$ compound [2].

4. Processes in the water layers can affect the ordering processes in the crystal lattice, that is, suppress the antiferromagnetic ordering as opposite to the ordering in superconductivity.

These four processes are the basic processes promoting the transition of $SrFe_2As_2$ and $FeTe_{0.8}S_{0.2}$ compounds into the superconducting state induced by water diffusion.

Stimulation of superconductivity in these compounds can take place and because of some other reasons.

When temperature decreases, coherence regions are expanded and in certain conditions can merge. All processes in such merged coherence regions occur simultaneously, with a single phase common for all the merged coherence area. The basic layers of ferriferous compound can be located in such merged coherence regions.

The synchronization of the processes occurring in coherent water layers and in ferriferous layers, apparently, also is one of the factors of stimulation of superconductivity in compounds $SrFe_2As_2$ and $FeTe_{0.8}S_{0.2}$.

References

[1] H. Hiramatsu, T. Katase, T. Kamiya et al., Water induced superconductivity in $SrFe_2As_2$. (2009) *Phys. Rev. B*. 80, 052501.

[2] Y. Mizuguchi, K. Deguchi, S. Tsuda, T. Yamaguchi, Y. Takano. Moisture-induced superconductivity in $FeTe_{0.8}S_{0.2}$. (2010) *Phys. Rev. B*. 81, 214510 [5 pages].

[3] K. Deguchi, Y. Mizuguchi, T. Ozaki et al., *Superconductivity in* $FeTe_{1-x}S_x$ *induced by alcohol*. arXiv:1008.0666.

[4] P. L. Alireza, Y. T. Chris Ko, J. Gillett et al., Superconductivity up to 29 K in $SrFe_2As_2$ and $BaFe_2As_2$ at high pressures. (2009) *J. Phys.: Condens., Matter*. 21. 012208 (4 p.).

[5] K. Kitagawa, N. Katayama, H. Gotou et al., Spontaneous formation of a superconducting and antiferromagnetic hybrid state in $SrFe_2As_2$ under high pressure. (2009) *Phys. Rev. Lett*. 103, 257002. K. Kitagawa, N. Katayama, H. Gotou, et al., *Spontaneous formation of a superconducting and antiferromagnetic hybrid state in* $SrFe_2As_2$ *under high pressure*. (2009) arXiv:0906.4740v3 [(2009) cond.-mat. Supr.-con.] 24 Nov.

[6] M. Kumar, M. Nicklas, A. Jesche, et al., Effect of pressure on the magnetostructural transition in $SrFe_2As_2$. (2008). *Phys. Rev. B*. 78, 184516.

[7] J. A. Izjumov, E. Z. Kumraev. A new a class of high-temperature superconductors in $FeAs$ - systems. (2008) *Success Phys. Sci.* (in Russian). V. 178, № 12, p. 1307-1334.

[8] Alaska Subedi, Lijun Zhang, D. J. Singh, M. H. Du. Density functional study of *FeS*, *FeSe*, and *FeTe*: Electronic structure, magnetism, phonons, and superconductivity. (2008) *Phys. Rev. B*. 78, 134514.

[9] R. H. Liu, T. Wu, H. Chen et al., A large iron isotope effect in $SmFeAsO_{1-x}F_x$ and $Ba_{1-x}K_xFe_2As_2$. (2009) *Nature* 459, 64-67.

[10] Parasharam M. Shirage, Kunihiro Kihou, Kiichi Miyazawa, et al., Inverse Iron Isotope Effect on the Transition Temperature of the *$(Ba,K)Fe_2As_2$* Superconductor (2009) *Phys. Rev. Lett.* 103, 257003.

[11] M. I. Sadovsky. High-temperature superconductivity in layered connections on a basis of iron. (2008) *Success Phys. Sci.* (in Russian). V. 178, № 12, p. 1243-12719.

[12] J. Zhao, D.-X. Yao, S. Li et al., Low energy spin waves and magnetic interactions in $SrFe_2As_2$. (2008) *Phys. Rev. Lett.* 101 167203.

[13] R. A. Ewings, T. G. Perring, R. I. Bewley. *High energy spin excitations in $BaFe_2As_2$.* (2008) arXiv:0808.2836v2 [cond.-mat. super-con] 20 Oct. 2008.

[14] L. S. Martseniuk. The properties of water in nanosistems, displayed at interaction with neutron and x-ray radiations. *Materials of the 10th international conference: "The interaction of radiations with a solid."* (2013). Minsk, Belarus, p. 58-61 (in Russian).

[15] J. Preparata. *QED Coherence in Condensed Water.* World Sci. Singapore. (1995). 236 p.

[16] E. Del Giudice. QED Coherence in Liquid Water and the Dynamics of the Organization of Water in Living Matter. 18-21 October (2007). INFN – Milano – Italy and IIB – Neuss – Germany. *Second Annual Conference on the Physics Chemistry and Biology of Water Mt. Snow*, Vermont, US.

[17] E. Del Giudice, P. R. Spinetti, A. Tedeschi. Water Dynamics at the Root of Metamorphosis in Living Organisms. (2010) *Water*, 2, p. 566-586. www.mdpi.com/ journal/water.

[18] G. H. Pollack, J. Clegg. Unexpected linkage between unstirred layers, exclusion zones, and water. (2008) *Phase Transitions in Cell Biology*; Pollack, G. H., Chin, W. C., Eds.; Springer Science and Business Media: Berlin, Germany, 143-152 p.

[19] V. J. Antonchenko, A. S. Davidov, V. V. Il'in. *Bases of the physics of water.* (1991). Kiev, "Naykova dumka," 672 p. (in Russian).

[20] D. I. Blokhintsev. *Bases of quantum mechanics* (1963). M. "Higher school." 619 p. (in Russian).

[21] I. V. Bargatin, B. A. Grishanin, V. N. Zadkov. The entangled quantum states of the atomic systems. (2001) *Successes Phys. Sci.* V. 171, № 6, p. 625-647 (in Russian).

[22] M. W. Ho. Super-Conducting Liquid Crystalline Water Aligned with Collagen Fibres in the Fascia as Acupuncture Meridians of Traditional Chinese Medicine. *Forum on Immunopathological Diseases and Therapeutics.* (2012). 3 (3-4), p. 221-236.

[23] M. W. Ho. Superconducting quantum coherent water in nanospace confirmed. (2012) *Sci. Soc.* 55: p. 48-51.

[24] A. I. Kolesnikov et al., Neutron Spectroscopy of Water in Carbon Nanotubes. (2004) *Phys. Rev. Lett.* 93, 035503.

[25] A. I. Kolesnikov et al., Anomalous Behavior of Proton Zero Point Motion in Water Confined in Carbon Nanotubes. (2006) *Phys. Rev. Lett.* 97, 247801.

[26] H. Ye, N. Naguib, Y. Gogotsi. TEM study of water in carbon nanotubes. (2004) *JEOL news.* 39, p. 2-7.

[27] N. Naguib, H. Ye, Y. Gogotsi, et al., Observation of water confined in nanometer channels of closed carbon nanotubes. (2004) *Nano Letters.* 4: 22, p. 37-43.

[28] Ho, M. W. First sighting of structured water. (2005) *Sci. Soc.* 28: p. 47-8.

[29] A. Kurcz, A. Capolupo, A. Beige, E. Del Giudice, G. Vitiello. Energy concentration in composite quantum systems. (2010) *Phys. Rev. A*, 81, 063821.

[30] N. I. Sinitsin, V. I. Petrosjan, V. A. Elkin, et al., The Special role of the system: «millimeter waves - the water environment in the nature». (1999) *Biomedical radio electronics* (in Russian). № 1, p. 3-21.

[31] A. V. Drozdov, T. I. Nagorskaya, S. V. Masyukevich, et al., The quant-mechanics aspects of effects of the weak magnetic fields on biological objects. (2001) *Biophysics*. V. 5, № 4, p. 740-749 (in Russian).

[32] B. P. Pjulman, A. P. Pjulman. *The quantum biochemistry*. (1965) M. "World," 654 p. (in Russian).

In: Superconductivity
Editor: Paulette Grant

ISBN: 978-1-63483-907-5

Chapter 3

POSSIBILITIES OF MANIPULATION OF CRITICAL TEMPERATURE IN SUPERCONDUCTIVITY USING NON-LINEAR CORRELATION MECHANISMS BETWEEN COOPER CARRIERS STIMULATED BY AN-HARMONIC LATTICE VIBRATIONS

*Nicolae A. Enaki**
Quantum Optics and Kinetic Processes Lab,
Institute of Applied Physics, Academy of Sciences of Moldova
Chisinau, MD, Republic of Moldova

Abstract

The possibility of selective excitation of the nonlinear lattice vibrations in high-temperature superconductors formed from layers or thin films (like in $YBa_2Cu_3O_{6.5}$, MgB_2, $La_{0.7}Sr_{0.3}MnO_3$, graphite layers, etc.) opens the new perspectives in the manipulations and control of quantum field in superconducting phase. In many cases the largest critical current is obtained in for thin films or multilayer structures as a function of the constraints of vibration or electron states. Here we note that the anisotropy of electronic and vibrational structures of the materials plays an important role in superconducting critical current or phase transition temperature. In this context the non-linear mechanisms of cooperative interaction between the layer current carriers through the anisotropy of the vibration states of thermostat are examined. One of such mechanism corresponds to the cooperative nonlinear interaction between the cooper pairs belonging to different layers in the single and two-phonon (or vibration) exchanges between the electrons. In this case the constant of cooperative interactions between the electrons increases as a function of temperature in the superconductivity phase. This effect is accompanied by the increasing of order parameter as a function of temperature, after which it begins to decrease till critical temperature.

The new ideas of non-linear interaction of carriers with lattice vibration in superconductivity phase of multilayer material are proposing taking into consideration the

*E-mail address: enakinicolae@yahoo.com; Website: http://www.qol.asm.md

anisotropy of lattice vibrations and nonlinear aspects of interaction of electrons with lattice vibrations. In the first case the coupling of longitudinal and transversal non-linear oscillation modes in multilayer superconductor materials substantially changes the dispersion law and the Debye frequency of the spectrum of the vibrational modes in quasi-harmonic approximation. As a mapping of this nonlinear coupling, it simulates the temperature increases of the correlation length between the carriers and the coupling integral between the them. The possibilities of the induced generation of coherent vibration modes in the transversal (or longitudinal) geometry of multilayer systems and the control of critical temperature are discussed in the contexts of modern experiments with Raman stimulation of these lattice vibrations in superconducting materials. The second aspect of nonlinear interaction is described by the Hamiltonian function, which takes into consideration two- and single quantum exchanges between the electrons through the lattice vibrations at finite temperature. The Cooper pairs and their coherent behavior are obtained, generalizing the method of elimination of lattice vibration operators for finite temperature. Such a vibration stimulation allows the control of solids in their electronic ground state and the reduced dissipation of direct lattice excitation makes it attractive for applications in functional material control. Many experimental results in this field shown that the electron interaction with non-linear lattice vibration is an important ingredient of the current scenario and can explain superconductivity in novel systems including high temperature superconducting.

1. Introduction

The coupling of electrons with lattice vibration in in normal metals and superconductors is of considerable importance for the understanding of the superconducting as well as transport properties. In most cases, the superconducting systems reduce their volume less than the normal ones. Consequently, anisotropy of multi-layer aspects of the superconducting phase cause stresses which are similar to stresses caused by the inhomogeneities due to the temperature. In many cases the $high - T_c$ superconductors have a layered structure, but the importance of this remains unclear. Some of these layers have chains of atoms with a local symmetry, which facilitates quasi-one-dimensional representation. Taking into consideration the interaction of the vibration along the layers with the vibrations perpendicular to the layer structure in this chapter we propose the re-normalization of Debye temperature Θ_D using the non-linear coupling between these two types of vibrations.

According to the description of superconductivity the critical temperature is an even more complex quantity. For simplicity, we express it within the Bardeen-Cooper-Schrieffer (BCS) [1] approximation $T_c = 0.85\Theta_D \exp(-1/\lambda)$, where λ is the electron-phonon coupling constant at $T = 0$. This expression suggests that by increasing the Debye frequency and electron-phonon coupling we may achieve the high critical temperature, T_c. In this chapter we propose that these two-parameters may be normalized with increasing of the temperature so that we can achieve the high temperature at the critical point.

The collective processes in condensed matter have many analogical proprieties with cooperative radiation effects in quantum optics. The role of quadratic coupled electron-phonon (e-ph) model of superconductivity was taken into consideration periodically in the literature [2]. The soliton (or bisoliton) model of superconductivity was proposed in Ref. [5] for an explanation of the superconductivity in organic quasi-one-dimensional conductors. The interest in the research of superconductivity was renewed in the high-T_c superconductivity in copper oxides (cuprates), proposed in Ref. [4]. For example the nonlinear (quadratic) electron-phonon coupling dominates in the $Bi_2Sr_2CaCu_2O_8$ system. At the microscopic level, any non-linear electron-phonon interaction can be described due to the coupling of electrons to a few phonon-complexes. Such coupling significantly influences the critical temperature. In our Ref. [6] the similar concept was proposed considering that in the non-linear interaction of Cooper-electrons with lattice vibrations take place in two-phonon exchanges through the virtual band of material, so that the coherence amplitude of Cooper-pairs (order parameter) firstly increases as a function of the temperature after which it decreases till zero value in the critical point T_c.

These concepts crystal structure of $high - T_c$ superconducting cuprates consists of $Cu - O$ - superconducting planes separated by thin insulating layers, between which pairs of superconducting electrons (the so-called Cooper pairs) can tunnel. The combination of the capacitive coupling of the layers and the possible excitations of the acoustic osculation between these planes gives rise to collective to modify the coupling between the Cooper-paired electrons. This selective excitation of infrared-active lattice vibrations to large amplitudes using mid-infrared light pulses is a powerful tool for the control of quantum matter [7, 8]. The similar ideas of the modification of the T_c are proposed in the Ref. [9] , where the authors proposed the designed and manufactured device for the automatic analysis of non-linear parameters of high-temperature superconducting ceramics $YBa_2Cu_3O_{7-x}$ in variable magnetic fields.

The well known polaron effect connected with re-normalizes the effective mass of the carriers and the phonon-mediated interaction between carriers leads to the formation of Cooper pairs, superconductivity in conventional low-temperature superconductors.

The discovery of the high-temperature The discovery of the high-temperature superconducting cuprates resulted in an interest in the role of lattice vibration, both in their normal state properties and the formation of the high transition temperature, which was originally assumed as arising from the strong interaction of electrons with lattice vibration. Many different scenarios, mostly based on magnetic fluctuation mechanisms of superconductivity have been developed [10] in order to understand the first principles of the coupling of electrons with lattice vibrations. More recent studies have shown, that interaction of electrons with lattice vibration in the CuO_2 layers are clearly seen in the normal and the superconducting properties of the high-T_c cuprates. It was recently proposed, that coupling of electrons to the in-plane vibration of the CuO_2 plane oxygen ions can be responsible for the structure of the normal state in this material [11].

Another material magnesium di-boron differs from ordinary metallic superconductors in several important ways, including the failure of conventional models to predict accurately its unusually high transition temperature, the effects of isotope substitution on the critical transition temperature, and its anomalous specific heat. A detailed examination of the energy associated with the formation of charge-carrying pairs, referred to as the superconducting energy gap, should clarify why MgB_2 is different. Some early experimental studies have indicated, that MgB_2 has multiple gaps [12],[13], but existed theoretical approaches [14],[15] have not explained from first principles the origin of these gaps and their effects. An important feature is that the electronic states dominated by orbitals in the boron plane couple strongly to specific the modes of lattice vibration, making pair formation favorable.

Up to now there has been no complete theoretical study on the influence of non-linearity (in the special cases the multi-phonon interaction and anharmonic of lattice vibration) on the superconducting transition temperature.

The collective processes of between the vibration modes and between the current carriers and lattice vibrations in condensed matter have many common properties with cooperative radiation effects of excited atoms in interaction with thermal bath of the electromagnetic field [16]. The BCS [1], theory of single phonon phase transition in superconductivity was in the center of attention of many authors [17]-[18]. Following the ideas developed in these papers, the new cooperative emission phenomenon for dipole-forbidden transitions of an inverted system of radiators can be observed in processes of two-photon spontaneous emission [19], [20]. Taking into consideration the ideas developed in these papers, the new theory of nonlinear superconductivity in which it is taking into account the single and two-phonon exchanges between the band electrons is proposed. Using the method of elimination of boson operators, it is obtained the temperature dependence of two-quantum exchange integrals between Cooper-electrons are obtained. This integral increase with increasing the temperature due to the fact that the thermal field of vibration lattice stimulates the two-quantum exchanges between the electrons. This is one of the main differences between the single and two phonon superconductivity theories. The non-linear (quadratic) electron-phonon coupling dominates in $Bi_2\ Sr_2CaCu_2O_8$ [11]. At the microscopic level any non-linear electron-phonon interaction can be described in terms of the coupling of electrons to a few phonon-complexes. Such coupling significantly influences the critical temperature [21].

The new methods of interaction of electronic subsystem with vibration modes of lattice are proposed in this chapter. In the section 2 we introduced the new creation and annihilation operators for the vibration modes which take into account nonlinear (quadratic) effect in the lattice vibration modes with potential energy $V(x) = m\omega^2\hat{x}^2/2 + A\hat{x}^4$. Taking into consideration the multi-layer aspects of $high - T_c$ superconductors we propose the coupling between the vibration modes in the plane of the layer with the vibration modes, propagated perpendicular to the layers. In this description the potential energy in the layers has the following coupling with the mode propagations along the layer: $V(x,y) = m_\perp\omega_\perp^2[\hat{x}^2 + \hat{y}^2]/2 + A[\hat{x}^2 + \hat{y}^2]\hat{z}^2$. The similar aspect has the potential energy of the vibration for the vibration modes along the Z axes perpendicular to the layers $V(z) == m_\parallel\omega_\parallel^2\hat{z}^2/2 + A[\hat{x}^2 + \hat{y}^2]\hat{z}^2$. In this case, we proposed the temperature re-normalization of the Debye frequency, ω_D, so that the critical temperature of superconducting Cooper paring in the (X,Y) planes of the layers may increase as the function of the sign of the coupling parameter A. The non-linear two-phonon transitions are represented in the language of these new operators in Sec. 3. After that in the section 4 we have studied the collective non-linear effects in interaction of electronic subsystem with the non-linear thermal bath of the lattice oscillators.

The interference between the single- and two-phonon exchange mechanisms in superconductivity is studied using a projection operator method developed in Sec. 4. It is observing the situation for which the temperature dependence part of exchange integral between the electrons can be larger than constant part described by single quantum coupling with lattice vibrations.

In the sections 4 and 5 it is demonstrated that for the positive anharmonic constant

A together with two-quantum transitions stimulate the increasing of the order parameter, which firstly increase with the increasing of the temperature, achieving its maximal value, after that decreases as in the traditional phase transition effects. In this case the correlations between first and second order electron-phonon interaction effects take place through nonlinear processes of lattice vibrations and two-quantum exchanges between the electrons.

2. Non-Linear Mechanisms of Electron Interaction with Anisotropic Lattice Vibrations

In this chapter we propose unify two mechanisms of non-linearity which appear between the Cooper-electrons in superconductivity. First of them is well known in the literature [6, 21] and is connected with lattice vibrations taking into consideration the multi-layers aspects of many high-temperature superconductors. A second mechanism depends on the electronic band structure and intrinsic symmetry of nonlinear electron-phonon interaction in multi-band structures in which the two-phonon transitions between the electronic states near the Fermi levels are similar to non-linear optical transition. Let us discuss in this section the nonlinear mechanism of electron-phonon interaction stimulated by lattice vibration.

According to the methods discussed in the solid state physics, the interaction Hamiltonian between of electrons and nuclei system is presented in the following way

$$\hat{H} = \sum_k \frac{\hat{P}_k^2}{2M} + \sum_k \frac{\hat{p}_k^2}{2m} + V(r,R),$$

where P_k, R, M and p_k, r, m are the momentum, position and mass of nuclei and electrons, respectively; $V(r,R)$ is the Coulomb energy of interaction between the electrons and nuclei described by the expression

$$V(r,R) = \sum_{j<k} \frac{e_j e_k}{|R_j - R_k|} + \sum_{j<k} \frac{e^2}{|r_j - r_k|} - \sum_{j<k} \frac{e_j e}{|R_j - r_k|}.$$

Here e_j and e are the electrical charge of the $j-th$ nucleus and electron respectively; R_j and r_k are the vector position of the \$k-th\$ electron and $j-th$ nucleus.

Let us firstly study the non-linear lattice vibration. In the adiabatic approximation we can introduce the displacement $\xi_j = R_j - R_j^0$. In this case we can regard the potential energy $U(R) \simeq \sum_{j<k} \frac{e_j e_k}{|R_j - R_k|}$ in Tailor series relatively its equilibrium state

$$U(R) = U(R_0) + \frac{1}{2} \sum_{l,j} \frac{\partial^2 U}{\partial R_l \partial R_j} \xi_j \xi_l + \frac{1}{4!} \sum_{l,j,m,n} \frac{\partial^4 U}{\partial R_l \partial R_j \partial R_m \partial R_n} \xi_j \xi_l \xi_m \xi_m. \quad (1)$$

In the single mode oscillation the potential energy (1) is reduced to the model system proposed in Ref. [14, 15], For $MgBr_2$ the displacement of the Boron atom in (xy) plane is described by the expression $E(u) = A_2 u^2 + A_4 u^4$, $A_2 = 9.4 eV/\AA$, $A_4 = 690.3 ev/\AA$. According to Hamiltonian ref the third order non-linearity in this symmetrical model is not taking into consideration. According to the Refs. [22, 23] in one dimension non-linear latices vibration this Hamiltonian is reduced to the simplest form

$$\hat{H}'_v = \sum_{j=1}^{N} \frac{\hat{\mathbf{P}}_j^2}{2M} + K\sum_{j}^{N} \frac{(\xi_j - \xi_{j-1})^2}{2} + A\sum_{j=1}^{N} (\xi_j - \xi_{j-1})^4 \quad (2)$$

where K and A are the harmonic and anharmonic coefficient in the direction ξ_j. The equation of motion is

$$\begin{aligned} M\ddot{\xi}_j &+ K(2\xi_j - \xi_{j+1} - \xi_{j-1}) \\ &\times [1 + 4(A/K)[(\xi_j - \xi_{j-1})^2 + (\xi_j - \xi_{j-1})(\xi_{j+1} - \xi_j) + (\xi_{j+1} - \xi_j)^2] = 0 \end{aligned}$$

In three dimensions models we can consider the nonlinear term of expression (2) as the renormalization of elastic constant K by other vibration modes of the crystal so that the potential energy can be represented in the following form

$$\hat{U} = \sum_{j,\alpha,\beta}^{N} \frac{(\xi_{j\beta} - \xi_{j-1})_\alpha^2}{2} \left[K + A(\xi_j - \xi_{j-1})_\beta^2\right], \quad (3)$$

where $(\xi_j - \xi_{j-1})^2 = \sum_{\alpha=x,y,z} (\xi_{\alpha j} - \xi_{\alpha j-1})^2$. This representation show, that the elasticity coefficient in the ξ-direction is normalized by the number of phonons in the other direction. This re-normalization field in the literature [24, 28] is named intensity coupling field. This re-normalization is attractive from the physical point of view in the discussion of the mutual influences between the vibration modes in multi-layer materials with anisotropy.

For example the material MgB_2 contains the layer structure in which the Boron atoms form honeycombed layers, and the magnesium atoms are located above the center of the hexagons in-between the boron planes may be substantially affected by anti-phase vibrations of the magnesium atoms in the Z direction. In this material the electronic states at the Fermi level, which are the highest occupied electronic states, are named σ - or π -bonding boron orbits (Figure 1). The σ - bonding states are confined in the boron planes and the electronic states of MgB_2 are strongly coupled with σ- vibrational modes of boron atoms in (X,Y) plane. Crystal structure of MgB_2 demonstrates that boron atoms form honeycomb planes, and magnesium atoms occupy the centers of the hexagons in-between boron planes. Thus, MgB_2 may be unique with partially occupied σ -bonding states in a layered structure. Because the charge distribution of the σ -bonding states is not symmetrical with respect to the in-plane positions of boron atoms, the σ -bonding states couple very strongly to the in-lyer vibration of boron atoms.

The discovery of high-temperature superconducting cuprates (La_2CuO_4, $YBa_2Cu_3O_6$, $YBa_2Cu_3O_7$, $Sr_{1-y}Nd_yCuO_2$ etc.), resulted in an interest in the role of the phonons, both in their normal state properties and in the formation of the high transition temperature, which was originally assumed as arising from strong electron-phonon coupling and nonlinearities, which may appear in such systems. Early first principles calculations of the electron-phonon coupling demonstrated that it cannot solely lead to T_c close to $100K$ [25]. More recent studies have shown that electron- phonon coupling in the CuO_2 layers can support, but not cause, d-wave gap superconductivity [26, 27]. Nevertheless, the effects of electron-phonon coupling are clearly seen in the normal, and the superconducting, properties of the high-T_c cuprates. It was recently proposed that coupling of electrons to the

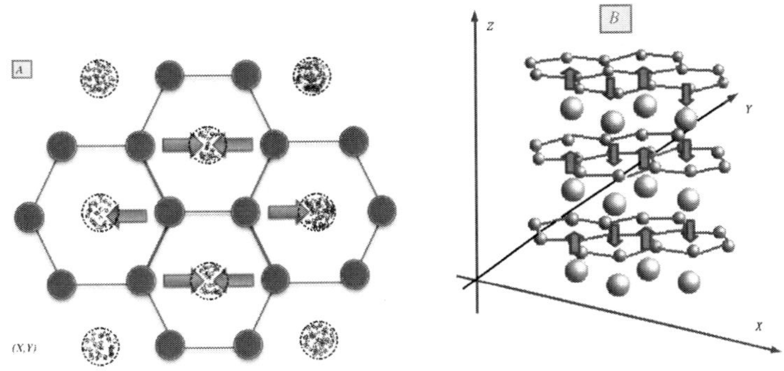

Figure 1. Crystal structure of MgB_2 . (A) Boron atoms form honeycomb planes, and magnesium atoms occupy the centres of the hexagons in-between boron planes. (B) Osculation of magnesium atom along the Z direction which may changes the dispersion law of the oscillation in (X,Y) plane represented in Figure A.

in-plane vibration of the CuO_2 plane, the oxygen ions can be responsible for the structure of the normal state of ARPES spectra, represented in Figure 2.

For example $YBa_2Cu_3O_6$ consists of CuO_2 layers and CuO_4 ribbons. The boundary of each layer is defined by planes of square planar CuO_4 units sharing 4 vertices. The layers can sometimes be slightly puckered. Perpendicular to these CuO_2-planes are CuO_4 ribbons sharing 2 vertices. The yttrium atoms are found between the CuO_2 planes, while the barium atoms are found between the CuO_4 ribbons and theCuO_2 planes. This structural feature is illustrated in the figure to the right.

The effective Debye frequencies of materials $YBa_2Cu_3O_6$ may substantially increase if the effective lattice force constant, $\kappa(\xi)$, of the oscillations in (X,Y) of CuO or CuO_2 planes on the increases substantially $\Omega^2 \sim \kappa/M$. As a result the critical temperature increases too, $T_c \sim (\kappa/M)^{1/2}$. The similar situation may take place in the planes of CuO_2 in which the Y and Ba atoms have the tendency to deform the bounds of O with Cu in CuO_2 planes. In other words the non-linear process can drastically change the frequencies of the oscillation in the one of direction (or plan) under the action of the nonlinear oscillation in another direction (such as Z direction described in Figure 2).

If we will name the atomic displacement by ξ-oscillation in (X,Y) plane and λ displacement in the z-direction of the oscillations, according to the above description of the vibration modes we may represent the non-linear coupling between layer (ξ_j) and vertical (λ_j) vibration modes by non-linear lattice vibration Hamiltonian

$$\begin{aligned}\hat{H}'_v &= \sum_{j=1}^{N} \frac{\hat{\mathbf{P}}_j^2}{2M_\perp} + \frac{\hat{\mathbf{P}}_j^2}{2M_\parallel} + K_\perp \sum_j^N \frac{(\xi_j - \xi_{j-1})^2}{2} \\ &+ K_\parallel \sum_j^N \frac{(\lambda_j - \lambda_{j-1})^2}{2} + \frac{A}{2} \sum_{\vec{j}(j_\perp, j_\parallel)} A(\xi_j - \xi_{j-1})^2(\lambda_j - \lambda_{j-1})^2. \end{aligned} \quad (4)$$

where K and A are the harmonic and an-harmonic coefficient in the direction ξ_j. Here we consider that the coupling takes place between the modes with the distortion in the parallel

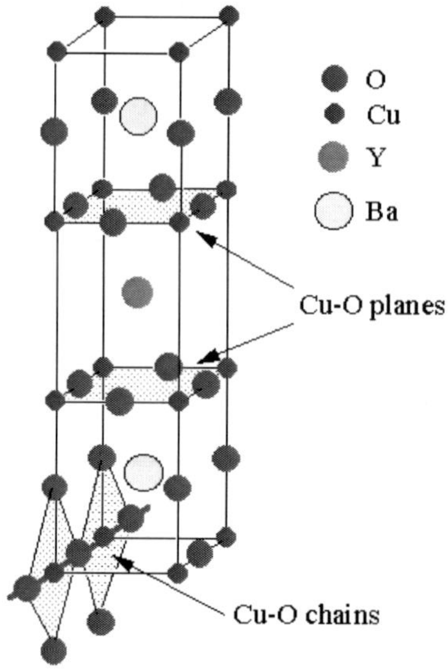

Figure 2. The high temperature superconductors are ceramic materials with layers of copper-oxide spaced by layers containing barium and other atoms. The yttrium compound is somewhat unique in that it has a regular crystal structure while the lanthanum version is classified as a solid solution. The yttrium compound is often called the $1-2-3$ superconductor because of the ratios of its constituents.

and perpendicular directions of the layer plane of the multilayer structures. The equations of motion for the atoms in this structure are

$$\begin{aligned} M_{\perp}\ddot{\xi}_j &+ (2\xi_j - \xi_{j+1} - \xi_{j-1})\{K_{\perp} + A\sum_{j_{\parallel}}(\lambda_j - \lambda_{j1})^2\} = 0, \\ M_{\parallel}\ddot{\lambda}_j &+ (2\lambda_j - \lambda_{j+1} - \lambda_{j-1})\{K_{\parallel} + A\sum_{j_{\perp}}(\xi_j - \xi_{j-1})^2\} = 0, \end{aligned} \tag{5}$$

where the sums on the index $j_{\parallel}$ and $j_{\perp}$ takes into consideration the nearest atoms, which enter in the chemical bounds with the j atom in vertical and the layer directions. In this anisotropic model of non-linear interaction of three-dimension vibration we may use the adiabatic time average on the period rapid oscillation of the one of subsystems (For example Boron Layer oscillations) described by the Hamiltonian (4). We see that such procedure of time renormalization of the elastic constant K of slow vibrations by rapid oscillations of other subsystem is possible and could be described by the potential function

$$\hat{U} = \sum_{j\beta}^{N} \frac{(\lambda_{j\beta} - \lambda_{j\beta-1})^2}{2}\left[K_{\parallel} + A\langle(\xi_j - \xi_{j-1})^2\rangle\right],$$

where time averaged sum, $\langle(\xi_j - \xi_{j-1})^2\rangle = \sum_{\alpha=x,y}\langle(\xi_{\alpha j} - \xi_{\alpha j-1})^2\rangle$, represents the re-normalization of the elasticity coefficient in the λ direction under the influence of rapid oscillations in the plane of the layers. This re-normalization field in the literature [24]-[28] is name intensity coupling field. In other words the effective lattice force constant becomes a function of amplitude and mean phonon numbers in the Z direction, $\kappa(\lambda) = K + A\sum_{j\beta}(\lambda_{j\beta} - \lambda_{j\beta-1})^2$.

In order to describe this field, let us firstly following the method of representation non-linear part of potential (1) through the annihilation $\hat{b}_q$ and creation $\hat{b}_q^\dagger$ operators of the phonon subsystem [29],[18]. According to this description we can represent the momentum $\mathbf{P}_j$ and displacements ξ_j in Fourier series

$$\begin{aligned} \hat{\xi}_j(\mathbf{R}_j) &= \sum_q \mathbf{g}_q \left[\hat{b}_q \exp(iq\mathbf{R}_j - i\omega_q t) + \hat{b}_q^\dagger \exp(-iq\mathbf{R}_j + i\omega_q t)\right]; \\ \hat{\mathbf{P}}_j(\mathbf{R}_j) &= -iM\sum_q \omega_q \mathbf{g}_q \left[\hat{b}_q \exp(iq\mathbf{R}_j) - \hat{b}_q^\dagger \exp(-iq\mathbf{R}_j)\right], \end{aligned} \tag{6}$$

where $\mathbf{g}_q = \sqrt{\hbar/(2NM\omega_q)}$. Using this representation [22] we can reduced the Hamiltonian (2) to the following form

$$\hat{H}_v = E_0 + \sum_q \hbar\omega_q \hat{b}_q^\dagger \hat{b}_q + \tilde{A}\hbar\omega_D \sum_{K,q_1,q_2} f_{q_1} f_{K-q_1} f_{q_2} f_{K-q_2} b_{q_1}^\dagger b_{K-q_1}^\dagger \hat{b}_{q_2} \hat{b}_{K-q_2}, \tag{7}$$

where $\omega_q \simeq cq\left(1 + \frac{4A}{\pi}\right)$, c is the sound velocity in the crystal, $f_q = sign(q)\sqrt{\sin(aq/2)}$, $\omega_D = 2\sqrt{\frac{K}{M}}$ is the Debye frequency, $\tilde{A} = 3\hbar A/\sqrt{K^3MN^2}$. The phonon operators obey the following commutation relations: $\left[\hat{b}_q; \hat{b}_{q_1}^\dagger\right]_- = \delta_{q,q_1}$; $\left[\hat{b}_q, \hat{b}_{q_1}\right]_- = 0$. In analogy with the with Wick-Bloch-Domincis [3] theorem we observe that the an-harmonic term change the sign when we take all wave vectors into account. In this situation it is better to approximate the operators $f_{q_1} f_{K-q_1} f_{q_2} f_{K-q_2} b_{q_1}^\dagger b_{K-q_1}^\dagger b_{q_2} b_{K-q_2}$ with expressions $\delta_{q_1,q_2} f_{q_1}^2 f_{K-q_1}^2 b_{q_1}^\dagger b_{K-q_1}^\dagger b_{q_1} b_{K-q_1}$ which don't change the signs. In this approximation the Hamiltonian (17) of lattice vibration becomes

$$\hat{H}_v = \sum_{\mathbf{q}} [\hbar\omega_q b_{\mathbf{q}}^\dagger \hat{b}_{\mathbf{q}} + \hbar\kappa_q \sum_{\mathbf{q}_1} f_{q_1}^2 \hat{b}_{\mathbf{q}}^\dagger \hat{b}_{\mathbf{q}_1}^\dagger \hat{b}_{\mathbf{q}_1} \hat{b}_{\mathbf{q}}], \tag{8}$$

where $\kappa_q = |\tilde{A}|\omega_D f_q^2$.

We can consider the negative value of nonlinear parameter $\kappa_q \leq 0$. In this situation we can introduce the collective operators of lattice subsystem in this nonlinear description where $\kappa_q = |\tilde{A}|\omega_D f_q^2$.

$$\hat{J}_q^+ = \frac{\hat{b}_q^\dagger}{f_q}\sqrt{\alpha - \sum_{q_1} f_{q_1}^2 \hat{n}_{q_1}}; \quad \hat{J}_q^- = \sqrt{\alpha - \sum_{q_1} f_{q_1}^2 \hat{n}_{q_1}}\frac{\hat{b}_q}{f_q}. \tag{9}$$

Here the parameter $\alpha = \omega_q/|\kappa_q|$ do not depend on the wave vector q; $\hat{n}_q = \hat{b}_q^\dagger \hat{b}_q$ is the phonon number operator in the vibration modes q. In this representation we can reduce these operators to $su(2)$ algebra. Indeed, commuting these operators, we obtain the new expression

$$[\hat{J}_q^+, \hat{J}_{q'}^-] = 2\delta_{q,q'}\hat{J}_{zq} + \frac{f_q}{f_{q'}}(1 - \delta_{q,q'})\hat{b}_q^\dagger \hat{b}_{q'}. \tag{10}$$

The $\hat{J}_{zq}$ operator

$$\hat{J}_{zq} = \frac{\hat{b}_q^\dagger \hat{b}_q}{2} - \frac{1}{2f_q^2}\left(\alpha - \sum_{q_1} f_{q_1}^2 \hat{n}_{q_1}\right),$$

together with the operators (9) belong the commutation relation for angular momentum components for the same wave vector q. This representation reduces the Hamiltonian of lattice vibration (8) to simplest form in comparison with (8)

$$\hat{H}_v = \sum_q \hbar f_q^2 \kappa_q \hat{J}_q^+ \hat{J}_q^-, \tag{11}$$

We observe that these commutation for different values of the q and q' can be neglected in comparison with $\hat{J}_{zq}$, when the number of phonons in other modes is larger than the phonon number in q mode $\sum_{q'} \hat{n}_{q'} f_{q'}^2 >> f_q^2 n_q$. Averaging on the states of other modes we obtain the frequency re-normalization of the q mode in quasi boson representation. Let us take an example connected with the multilayer structure of superconducting material described above. The modification of the vibration modes of the (X,Y) layers under the influence of the vibration in Z direction can be easily obtained from the expressions (9) and (13). Representing the operators in the Z direction by the expressions $\hat{B}_{q_\parallel}^\dagger$ and $\hat{B}_{q_\parallel}$, we obtain the re-normalization coefficients for the annihilation and generation operators of (X,Y)-vibration modes $\hat{J}_{q_\perp} = \sqrt{\alpha - f^2 \hat{B}_{q_\parallel}^\dagger \hat{B}_{q_\parallel}} \hat{b}_{q_\perp}/f$ and $\hat{J}_{q_\perp}^\dagger = \hat{b}_{q_\perp}^\dagger \sqrt{\alpha - f^2 \hat{B}_{q_\parallel}^\dagger \hat{B}_{q_\parallel}}/f$. If we consider that the vibration modes in the Z direction have lower frequencies and in this situation the occupation phonon number becomes overcrowded for lower temperature than in the (X,Y) layer vibration modes (see Figure 1). This is possible due to the fact that the Debye frequency in the Z direction is less than in (X,Y) layer vibration modes due to the inverse proportionality of these frequencies to the squirt of the atomic masses of Mg and B, respectively. The atomic mass estimation of Mg and B is 24.312 and 10.811 *amu* and respectively, we obtain the inequality: $\Omega_{D_\parallel} < \Omega_{D_\perp}$, where $\Omega_{D_\parallel} \sim 1/\sqrt{M_{Mg}}$; $\Omega_{D_\perp} \sim 1/\sqrt{M_B}$. In this situation we may average the creation and annihilation operators for the vibrations in the layers on the quantum states of low frequency vibration modes along the Z direction. Introducing the mean phonon number in this direction $N_\parallel = \langle \hat{B}_{q_\parallel}^\dagger \hat{B}_{q_\parallel} \rangle$, we obtain the temperature decreasing of the renormalized Debye frequency in the (X,Y) plane: $\tilde{\Omega}_{D_\perp} \simeq \Omega_{D_\perp} - f^2 \kappa_\parallel N_\parallel$.

From the commutation relations 10) it follows that in the case of intrinsic non-linearity of the vibration modes, we can reduce this commutation relation to the operators of $su(2)$ algebra algebra [30]

$$\begin{aligned} \hat{J}_q^+ &= \frac{\hat{b}_q^\dagger}{f_q} \sqrt{\alpha - f_q^2 \hat{n}_q}; \quad \hat{J}_q^- = \sqrt{\alpha - f_q^2 \hat{n}_q} \frac{\hat{b}_q}{f_q} \\ \hat{J}_{zq} &= -\frac{\alpha}{2 f_q^2} + \hat{b}_q^\dagger \hat{b}_q. \end{aligned}$$

which exactly satisfy the the commutation relation (12). In this representation we can reduce the commutation relation between the new operators $\hat{J}_q^+$, $\hat{J}_q^-$ and $\hat{J}_{zq}$ to the $su(2)$ algebra operators of angular momentum

$$[\hat{J}_{\mathbf{q}}^+, \hat{J}_{\mathbf{q}'}^-] = 2\delta_{q,q'} \hat{J}_{q'}^z; \quad [\hat{J}_q^z, \hat{J}_{\mathbf{q}'}^\pm] = \pm \delta_{q,q'} \hat{J}_{\mathbf{q}}^\pm. \tag{12}$$

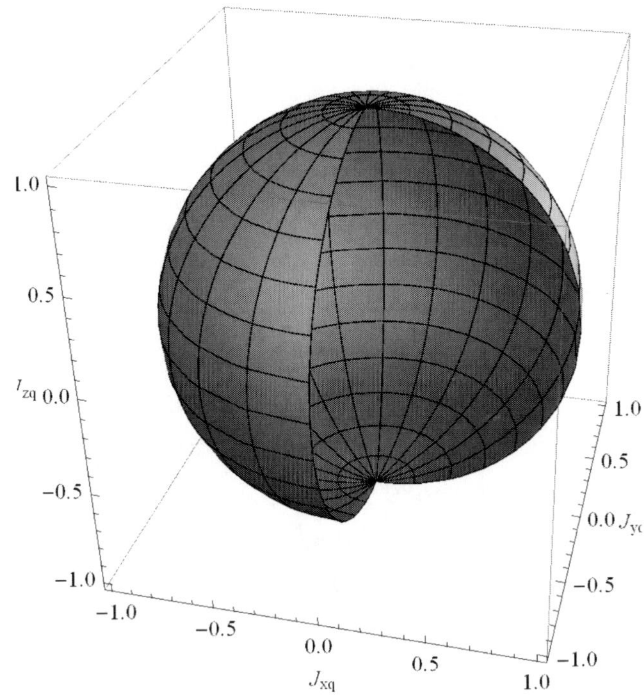

Figure 3. The Kazimir vector $j_q(j_q+1) = J_{qx}^2 + J_{qy}^2 + J_{qz}^2$ for given su(2) algebra for given wave vector q and $2j_q = \alpha/f_q^2$, where $J_{qx} = J_q^+ + J_q^-/2$ and $J_{qy} = i(J_q^+ - J_q^-)/2$.

This representation of the vibration modes through the operators of $su(2)$ for the given wave vector q it is described by the conservation of the Kazimir vector $J^2 = \hat{J}_{zq}^2 + \{\hat{J}_q^+\hat{J}_q^- + \hat{J}_q^-\hat{J}_q^+\}/2$, which has the best geometric presentation plotted in Figure 3. We can find the action of this operator to the Hilbert vectors $|m_q, j_q\rangle$ of angular momentum $su(2)$ algebra

$$\begin{aligned}
\hat{J}^2|m_q, j_q\rangle &= j_q(j_q+1), \\
\hat{J}_{zq}|m_q, j_q\rangle &= m_q|m_q, j_q\rangle, \\
\hat{J}_q^+|m_q, j_q\rangle &= \sqrt{(j_q - m_q)((j_q + m_q + 1)}|m_q + 1, j_q\rangle, \\
\hat{J}_q^-|m_q, j_q\rangle &= \sqrt{(j_q - m_q + 1)((j_q + m_q)}|m_q - 1, j_q\rangle
\end{aligned}$$

where, $m_q = -j_q, -j_q+1, ..., j_q$. For zero temperature all modes are in the ground state $|-j_q, j_q\rangle$, which corresponds to the zero value of the number of phonons $\hat{b}_q^\dagger\hat{b}_q|-j_q, j_q\rangle = 0$. In this case the parameter $2j_q = \alpha/f_q^2 = (N\omega_q/(|\tilde{A}|\omega_D f_q^4)$.

The increasing of the re-normalisation Debye frequency can be obtained for the positive sign of anharmonic parameter, $\kappa_q > 0$. In this case we can introduce the new operators

$$\hat{I}_q^+ = \frac{\hat{b}_q^\dagger}{f_q}\sqrt{\alpha + \sum_{q_1} f_{q_1}^2 \hat{n}_{q_1}}; \quad \hat{I}_q^- = \sqrt{\alpha + \sum_{q_1} f_{q_1}^2 \hat{n}_{q_1}}\frac{\hat{b}_q}{f_q}, \tag{13}$$

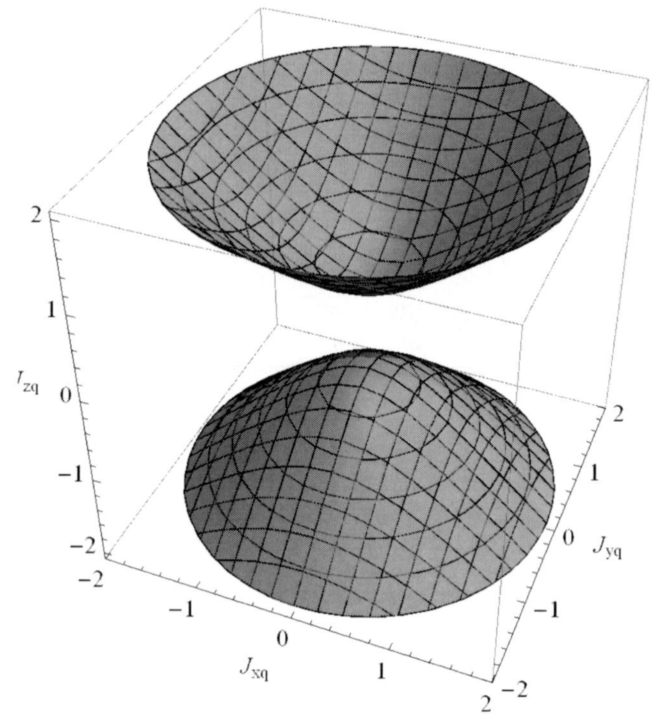

Figure 4. The rotation hyperboloid $j_q(j_q-1) = I_{qz}^2 - I_{qx}^2 - I_{qy}^2$ for following parameters of the system $2j_q = \alpha/f_q^2$, where $I_{qx} = I_q^+ + I_q^-/2$ and $I_{qy} = i(I_q^+ - I_q^-)/2$.

which for same value of wave-vector q are reduced to the operators of $su(1,1)$ algebra

$$[\hat{I}_q^+, \hat{I}_{q'}^-] = -2\delta_{q,q'}\hat{I}_{zq} + \frac{f_q}{f_{q'}}(1-\delta_{q,q'})\hat{b}_q^\dagger \hat{b}_{q'}, \quad (14)$$

where

$$\hat{I}_q^z = \frac{\hat{b}_q^\dagger \hat{b}_q}{2} + \frac{1}{2f_q^2}\left(\alpha + \sum_{q_1} f_{q_1}^2 \hat{n}_{q_1}\right)$$

. We may neglect the second term in the right hand site of the expression (14), when $\sum_{q'} \hat{n}_{q'} f_{q'}^2 >> \hat{n}_q f_q^2$. This corresponds to the situation when the Bose propriety of new operators is possible under the re-normalization of the frequency of the q mode in the process of interaction with other modes from the system. This situation can be applied to the High Tc multi-layer superconducting materials, in which the plane vibration is normalized by lower frequency vibrations in perpendicular direction to the layers. Due to mass differences between the atoms taking part in such vibrations we can use the similar representation for planer and perpendicular vibration modes as in $su(2)$ algebra :$\hat{I}_{q_\perp} = \sqrt{\alpha + f^2 \hat{B}_{q_\parallel}^\dagger \hat{B}_{q_\parallel}}\hat{b}_{q_\perp}/f$ and $\hat{I}_{q_\perp}^\dagger = \hat{b}_{q_\perp}^\dagger \sqrt{\alpha + f^2 \hat{B}_{q_\parallel}^\dagger \hat{B}_{q_\parallel}}/f$. In this case the re-normalization of the Debay frequency for this layer vibration becomes $\tilde{\Omega}_{D_\perp} \simeq \Omega_{D_\perp} + f^2 \kappa_\parallel N_\parallel$. The large value of the Mg-atom in comparison with the B one gives us the possibility to consider that the value of the Debye frequency $\Omega_{D_\parallel}$ in the modes propagated in the perpendicular direction of the layers is less than one in the layers. In this situation the average number of thermal fonons in the layer modes with increasing the temperature may remain smaller than one $N_{q_\perp} = 1/[\exp(\hbar\omega_{q_\perp}/k_B T) - 1] << 1$ and of course can be less than the number of the phonons in the modes propagating along the Z axis ($\hbar\omega_{q_\perp} > \hbar\omega_{q_\parallel}$).

$$[\hat{I}^+_{\mathbf{q}}, \hat{I}^-_{\mathbf{q}'}] \simeq -2\hat{I}^z_q \delta_{q,q'}; \quad [\hat{I}^z_{q'}, \hat{I}^\pm_{\mathbf{q}}] \simeq \pm \hat{I}^\pm_{\mathbf{q}} \delta_{q,q'},$$

which belong to $su(1,1)$ algebra. In the similar approximation as in the $su(2)$ symmetry we observe, that for the same values of the q the commutation relations (10) and (14) are reduced to exact operator representations for $su(1,1)$ operator algebras. In the case, when $\kappa_q > 0$, the Hamiltonian of the lattice vibration becomes

$$\hat{H}_v = \sum_q \hbar f_q^2 \kappa_q \hat{I}^+_q \hat{I}^-_q. \tag{15}$$

For positive an-harmonic parameter the new Kazimir vector $I^2 = \hat{I}^2_z - (\hat{I}^+_q \hat{I}^-_q + \hat{I}^-_q \hat{I}^+_q)/2$ commutes with the Hamiltonian (15) (see Figure 4). For intrinsic non-linear modes, we can show that the above commutation relation is exactly for $su(1,1)$ algebra operators

$$\begin{aligned} \hat{I}^+_q &= \frac{\hat{b}^\dagger_q}{f_q}\sqrt{\alpha + f_q^2 \hat{n}_q}; \quad \hat{I}^-_q = \sqrt{\alpha + f_q^2 \hat{n}_q}\frac{\hat{b}_q}{f_q}; \\ \hat{I}^z_q &= \frac{\alpha}{2f_q^2} + \hat{b}^\dagger_q \hat{b}_q. \end{aligned} \tag{16}$$

The discrete energetically states of $su(1,1)$ algebra are described by complete set of the state vectors $|m_q, j_q\rangle$ for wave functions. The actions of the above operators are represented by the relations:

$$\begin{aligned} \hat{I}^2|m_q, j_q\rangle &= j_q(j_q - 1)|m_q, j_q\rangle, \\ \hat{I}_{zq}|m_q, j_q\rangle &= (m_q + j_q)|m_q, j_q\rangle, \\ \hat{I}^+_q|m_q, j_q\rangle &= \sqrt{(m_q+1)(2j+m)}|m_q+1, j_q\rangle, \\ \hat{I}^-_q|m_q, j_q\rangle &= \sqrt{m_q(2j+m-1)}|m_q-1, j_q\rangle. \end{aligned}$$

from which follows that the ground state of the vibration mode q is $\hat{I}^-_q|0, j_q\rangle = 0$. Here j_q is the Bargmann index [31] and m_q is the positive integer number. The set of vectors orthogonal vectors, $|m_q, j_q\rangle$, $m = 0, 1, 2, \ldots$, described in the upper sheet of rotation hyperboloid (see Figure 4) becomes complete $\sum_{m_q} |m_q, j_q\rangle\langle m_q, j| = 1$ and orthogonal $\langle m'_q, j|m_q, j_q\rangle = \delta_{m'_q, m_q}$for given q.

Let us unify the Hamiltonian forms (11) and (15) in the new representation introducing the operators $\hat{K}^+_q$, $\hat{K}^-_q$ and $\hat{K}_{zq}$ for both algebras $su(2)$ and $su(1,1)$. For the both signs of the parameter A we can introduce the commutation relations for these operators depending on the sign of non-linear parameter $[\hat{K}^+_q, \hat{K}^-_q] = -\sigma\hat{K}_{zq}$ and $[\hat{K}_{zq}, \hat{K}^\pm_q] = \pm\hat{K}^\pm_q$ with conserved vector $K^2 = \hat{K}^2_{zq} - \sigma(\hat{K}^+_q\hat{K}^-_q + \hat{K}^-_q\hat{K}^+_q)/2$. Here the sign of parameter σ coincide with the sign of nonlinearity. In this representation the expressions (11) and (15) of nonlinear vibration Hamiltonian can be unified from which follows that the ground state of the vibration mode q is $\hat{I}_q^-|0, j_q\rangle = 0$. Here j_q is the Bargmann index [31] and m_q is the positive integer number. The set of vectors orthogonal vectors, $|m_q, j_q\rangle$, $m = 0, 1, 2, ...$, described in the upper sheet of rotation hyperboloid (see Figure 4) becomes complete $\sum_{m_q} |m_q, j_q\rangle\langle m_q, j| = 1$ and orthogonal $\langle m'_q, j|m_q, j_q\rangle = \delta_{m'_q,m_q}$for given q.

Let us unify the Hamiltonian forms (11) and (15) in the new representation introducing the operators $\hat{K}_q^+$, $\hat{K}_q^-$ and $\hat{K}_{zq}$ for both algebras $su(2)$ and $su(1,1)$. For the both signs of the parameter A we can introduce the commutation relations for these operators depending on the sign of non-linear parameter $[\hat{K}_q^+, \hat{K}_q^-] = -\sigma\hat{K}_{zq}$ and $[\hat{K}_{zq}, \hat{K}_q^\pm] = \pm\hat{K}_q^\pm$ with conserved vector $K^2 = \hat{K}_{zq}^2 - \sigma(\hat{K}_q^+\hat{K}_q^- + \hat{K}_q^-\hat{K}_q^+)/2$. Here the sign of parameter σ coincides with the sign of nonlinearity. In this representation the expressions 11) and (15) of nonlinear vibration Hamiltonian can be unified

$$\hat{H}_v = \sum_q \varepsilon_q \hat{K}_q^+ \hat{K}_q^+. \tag{17}$$

Here $\varepsilon_q = \hbar f_q^2 \kappa_q$. The nonlinear displacement $\mathbf{u}(x)$ can be represented through new operators like expressions (6)

$$\hat{\mathbf{u}}(\mathbf{x},t) = \sum_q \mathbf{q}_q \left[\hat{K}_q^-(t)\exp(i\mathbf{q}\mathbf{x}) + \hat{K}_q^\dagger(t)\exp(-i\mathbf{q}\mathbf{x})\right], \tag{18}$$

where $\mathbf{q}_q = \mathbf{g}_q f_q/\sqrt{\alpha}$.

Below we found the possibility to use these new operators in the interaction representation of electron subsystem with lattice vibration. Following the well known two interaction models, described in the literature [29, 32], we observe that one of them corresponds to the dipole deformed structure of the atomic orbital in the process of lattice vibrations. The second approach corresponds to the non-deformation dipole structures of the atoms in the lattice. In the second case it is considered that the periodical potential $V_e(r,R)$ of electrons are disturbed only to the displacement of the lattice, so the non-periodical potential energy can be defined by the expression $V_I(r,R) = V_e(\mathbf{r}+\mathbf{u},R) - V_e(r,R)$. First model corresponds to the situation, when the electrons and lattice vibration is considered in the framework of continual model. The polarization oscillations we must consider in other way like in the nonlinear optics. Indeed, following the procedure of deduction of Frohlich Hamiltonian, we observe, that the interaction of the charge density $\hat{\rho}(x) = e\hat{\Psi}^+(x)\hat{\Psi}(x)$ in the point $\mathbf{x}$ with the dipole density $\hat{\mathscr{P}}(\mathbf{x}')$ formed from the deformation of electronic orbital and nuclei charge in point $\mathbf{x}'$ is given by the expression

$$\hat{\rho}(x)(\frac{(\mathbf{x}'-\mathbf{x})}{|\mathbf{x}-\mathbf{x}'|^3}, \vec{P}(\mathbf{x}'))d^3xd^3x'.$$

Here $\hat{\Psi}^+(x) = V^{-0.5}\sum_p u_p^* \hat{a}_p^\dagger \exp[-i\mathbf{p}\mathbf{x}]$ and $\hat{\Psi}(x)$ are the operators of creation and annihilation of the electron in the point x which satisfy the commutation relation $[\hat{\Psi}(\mathbf{x},t), \hat{\Psi}^+(\mathbf{x}',t)] = \delta^3(\mathbf{x}-\mathbf{x}')$; $\hat{a}_p^\dagger$ and $\hat{a}_p$ are the creation and annihilation operators of the electrons with the momentum p, $u_p^*(x)$ is the Bloch amplitude of the $p-$ state, V is the volume of the crystal. In this case the Interaction Hamiltonian between the electrons and the polarization is

$$\hat{H}_I = -\int d^3x \hat{\mathscr{P}}(x)\hat{\mathscr{D}}(x), \tag{19}$$

where the electrical displacement $\hat{\mathscr{D}}(x)$ is

$$\hat{\mathscr{D}}(x) = \int d^3x' \hat{\rho}(x') \frac{(\mathbf{x}-\mathbf{x}')}{|\mathbf{x}-\mathbf{x}'|^3} = -\frac{\partial}{\partial \mathbf{x}} \int d^3x' \frac{\hat{\rho}(x')}{|\mathbf{x}-\mathbf{x}'|}.$$

Following this method we can regard the lattice vibration in interaction with electron subsystem with the Lagrange function $\hat{L} = \int d^3x \hat{\mathscr{L}}(\mathbf{x}) = \hat{L}_0 + \hat{L}_I$ the density of which can be represented through the free and interaction parts

$$\begin{aligned} \hat{\mathscr{L}}_0(\mathbf{x},t) &= \frac{\gamma}{2} \dot{\hat{\mathscr{P}}}^2(\mathbf{x}) - g\frac{(\nabla \hat{\mathscr{P}}(\mathbf{x}))^2}{2} - \eta(\nabla \hat{\mathscr{P}}(\mathbf{x}))^4(\mathbf{x}), \\ \hat{\mathscr{L}}_I(\mathbf{x},t) &= \hat{\mathscr{P}}(\mathbf{x})\hat{\mathscr{D}}(\mathbf{x}). \end{aligned}$$

Here the constants $\gamma \sim aM/q^2$, $g \sim aK/q^2$ and $\eta \sim a^3K/q^3$ play the similar role as in Hamiltonian (17); a and q are lattice constant and nuclei charge respectively. The Euler-Lagrange equation [33], [34]

$$\partial_\mu \frac{\partial \hat{\mathscr{L}}(x)}{\partial(\partial_\mu \hat{\mathscr{P}}_\alpha(x))} = \frac{\partial \hat{\mathscr{L}}(x)}{\partial \hat{\mathscr{P}}_\alpha(x)},$$

gives us the possibility to found the equation of motion of polarization

$$\gamma \ddot{\hat{\mathscr{P}}}(\mathbf{x}) - g \nabla_i [1+4(\eta/g)(\nabla \hat{\mathscr{P}}(\mathbf{x}))^2] \nabla_i \hat{\mathscr{P}}_\alpha(\mathbf{x}) = \int \frac{(\mathbf{x}-\mathbf{x}')_\alpha}{|\mathbf{x}-\mathbf{x}'|^3} \rho(x')d^3x' = \hat{\mathscr{D}}_\alpha(x). \quad (20)$$

Here $\{\partial_\mu\} = \{\frac{\partial}{\partial t}, \frac{\partial}{\partial x_1}, \frac{\partial}{\partial x_2}, \frac{\partial}{\partial x_3}\}$. Considering that the process of the polarization oscillations is damped, we can neglect the first term from the equation (20) $M\ddot{\hat{\mathscr{P}}}(\mathbf{x}) \to 0$. In this situation the approximative connection between the polarization $\hat{\mathscr{P}}(\mathbf{x})$ and electrical displacement $\hat{\mathscr{D}}(x)$ is given by the expression

$$-g \nabla_i [1+4(\eta/g)(\nabla \hat{\mathscr{P}}(\mathbf{x}))^2] \nabla_i \hat{\mathscr{P}}_\alpha(\mathbf{x}) \simeq -\nabla_\alpha \int d^3x' \frac{\rho(x')}{|\mathbf{x}-\mathbf{x}'|}.$$

It follows that this connection can be regarded in simplest form

$$g \nabla_i \hat{\mathscr{P}}_\alpha(\mathbf{x})[1+4(\eta/g)(\nabla \hat{\mathscr{P}}(\mathbf{x}))^2] - \delta_{i,\alpha} \int d^3x' \frac{\hat{\rho}(x')}{|\mathbf{x}-\mathbf{x}'|} = 0.$$

Let us introduce this relation in interaction Hamiltonian (19).

$$\hat{H}_I = \int [1+4(\eta/g)(\nabla, \hat{\mathscr{P}}(\mathbf{x}))^2](\nabla, \hat{\mathscr{P}}(\mathbf{x}))^2 d^3x$$

According to the electrostatic, we can use instead of the polarization $(\nabla, \mathscr{P}(\mathbf{x}))$ the following representation through the displacement $\hat{\mathscr{D}}$

$$4\pi div \hat{\mathscr{P}}(\mathbf{x}) = \left(1 - \frac{1}{\varepsilon}\right) div \hat{\mathscr{D}}(\mathbf{x}) = \left(1 - \frac{1}{\varepsilon}\right) 4\pi \hat{\rho}(x).$$

In this situation the interaction Hamiltonian takes the form

$$\hat{H}_I = -g\left(1-\frac{1}{\varepsilon}\right)\int \hat{\rho}(x)[1+4(\eta/g)(\nabla\hat{\mathscr{P}}(\mathbf{x}))^2](\nabla,\hat{\mathscr{P}}(\mathbf{x}))d^3x \tag{21}$$

The connection between the polarization $\hat{\mathscr{P}}(x)$ and non-linear deformation can be found if we approximate the interaction part of the Hamiltonian (21) with the expression of the left hand site of the steady-state equation (20). In one-dimensional model we have

$$g[1+4(\eta/g)(\nabla\hat{\mathscr{P}}(\mathbf{x}))^2]\nabla\hat{\mathscr{P}}(\mathbf{x}) \to a\left(K\frac{\partial\xi(x_j)}{\partial x_j}+4Aa^2\left(\frac{\partial\xi(x_j)}{\partial x_j}\right)^3\right),$$

This representation suggests us that in the case of three-dimension isotropic model we must introduce instead of a linear deformation tensor the divergence of renormalized deformation $div\mathbf{u}(x)$

$$\begin{aligned}
&\int d^3x\hat{\Psi}^+(x)\hat{\Psi}(x)div\mathbf{u}(x) \to i\int d^3x\hat{\Psi}^+(x)\hat{\Psi}(x)\sum_q(\mathbf{q},\mathbf{q}_q)\\
&\times\left[\hat{K}_q^-\exp(i\mathbf{qx})-\hat{K}_q^\dagger\exp(-i\mathbf{qx})\right]\\
= \; & i\sum_{p,q}(\mathbf{q},\mathbf{q}_q)\{\hat{a}^\dagger_{p+q}(t)\hat{a}_p(t)\hat{K}_q^-(t)-\hat{a}^\dagger_p(t)\hat{a}_{p+q}(t)\hat{K}_q^+(t)\}.
\end{aligned}$$

Introducing this representation in the expression (21), we observe that in the interaction Hamiltonian part the interaction constant $g\left(1-\frac{1}{\varepsilon}\right)$ is not defined. In order to solve this problem, we propose to use by the deformation approach [29]. According to this, let us express the real deformation tensor $\eta_{ij}(\xi_j)$ through the nonlinear displacement of the dipoles $\mathbf{u}(x_j)$

$$\eta_{ij}(\xi_j) = \frac{1}{2}\left(\frac{\partial u_j}{\partial x_i}+\frac{\partial u_i}{\partial x_j}\right).$$

The relative variation of the volume in the process of the deformation can be described by the expression $\rho \equiv (\delta\widetilde{V}-\delta V)/\delta V$ where $\delta\widetilde{V} = (1+\eta_{11})(1+\eta_{22})(1+\eta_{33})\delta V \simeq (1+\sum_i\eta_{ii})\delta V$. In other words the relative changes in the volume $\rho = Sp\eta = div\mathbf{u}(\mathbf{r})$. For small deformations, the energy of band electrons near the Fermi level is represented through the tensor of deformations

$$E(\mathbf{k},\eta(\xi)) = E(\mathbf{k}) - \sigma Sp\{\eta(\xi)\}.$$

where the parameter σ can be found if you consider that the variation of the Fermi energy $E_f = \hbar^2(3\pi^2N)^{2/3}/(2m^*V^{2/3})$ is represented through the $\eta-$ tensor

$$\delta E_f = -\frac{2}{3}E_f\frac{\delta V}{V} = -\frac{2}{3}E_fSp\{\eta\}.$$

Here N and m^* are the number of electrons and effective mass respectively. According to this description the parameter ς can be connected with the position of the Fermi level $\varsigma = 2E_f/3$. For ideal crystals with dispersion law $E(k) = \hbar^2k^2(2m^*)^{-1}$ the electron-phonon interaction energy is represented through the deformation tensor according to the expression $H_I = E(\mathbf{k}, \eta(\xi)) - E(\mathbf{k})$. In this case we have $H_I(\mathbf{x}) = -\varsigma Sp\{\eta)\} = -\varsigma(\nabla, \mathbf{u}(\mathbf{x}))$. Introducing the Fourier transformation (6), we can find following interaction Hamiltonian

$$\hat{H}_I = -i\sum_{\mathbf{k},\mathbf{q}} Q_1(q)\{\hat{a}^\dagger_{k+q} a_k \hat{K}^-_{\mathbf{q}} - a^\dagger_{\mathbf{k}} a_{\mathbf{k}+\mathbf{q}} \hat{K}^\dagger_{\mathbf{q}}\}, \tag{22}$$

Where the interaction constant is $Q_1(q) = \varsigma(\tilde{\mathbf{q}}, \mathbf{q}_q)$.

We observe, that in the case of the material $Bi_2Sr_2CaCu_2O_8$, the vibration mode ξ_j along the plan, where is situated CuO_2 [11] or along the cuprous layers of $MgCu_2$, the frequencies are evidently intensity dependent. Indeed, using the decomposition (ref eq: Fourier) for ξ_j and λ_l displacement along the plan where are situated CuO_2 and perpendicular to this plane we can establish that the frequencies.

$$\begin{aligned}
\omega^\perp_q &\sim 2v_1\sqrt{1 + (A/K_\perp)\sum_{q'} G^2_q(\hat{B}^\dagger_{q'}\hat{B}_{q'} + \hat{B}_{q'}\hat{B}^\dagger_{q'})\sin q/2}, \\
\omega^\parallel_q &\sim 2v_2\sqrt{1 + (A/K_\parallel)\sum_{q'} g^2_{q'}(\hat{b}^\dagger_{q'}\hat{b}_{q'} + \hat{b}_{q'}\hat{b}^\dagger_{q'})\sin q/2}.
\end{aligned} \tag{23}$$

so that the proposed displacement $\mathbf{u}(\mathbf{x})$ and its Fourier transformation (18) take into consideration this renormalization of frequencies by other thermal modes. Indeed, using the commutation of the operators $\hat{K}^\pm_q$ with Hamiltonian (17) we can observe that the Heisenberg equation for new annihilation or creation $\hat{K}^\pm$ operator takes into consideration the character of the nonlinear vibration through the Z- component of $su(2)$ and $su(1,1)$ group operators

$$\frac{d\hat{K}^\pm_q(t)}{dt} = \pm 2\sigma i \varepsilon_q \hat{K}_{zq}(t)\hat{K}^\pm_q(t),$$

where $2\sigma\varepsilon_q\hat{K}_{zq}(t)$ in the thermal bath can be approximated with its mean value $\tilde{\omega}_q \simeq \omega_q + \sigma\kappa_q\sum_{q'} f^2_{q'}\langle\hat{b}^\dagger_{q'}\hat{b}_{q'}\rangle(1+\delta_{q,q'})$. We can mention that the temperature re-normalization of oscillation frequencies is not so simple procedure. In order to consider that $\mathbf{u}(\mathbf{x}',\mathbf{t})$ is the displacement in one direction of the crystal re-normalized by other vibration modes, we must introduce the momentum of this non-linear field

$$\vec{\mathscr{P}}(\mathbf{x},t) = -iM\sum_q q\mathbf{q}_q\left[\hat{K}^-_q\exp(i\mathbf{q}\mathbf{x}) - \hat{K}^\dagger_q\exp(-i\mathbf{q}\mathbf{x})\right].$$

The mean value of the commutator of this operator with the displacement $\hat{\mathbf{u}}(\mathbf{x},\mathbf{t})$ must have the proprieties near propriety of Boson field $\langle[\hat{\mathbf{P}}(\mathbf{x},\mathbf{t}), \hat{\mathbf{u}}(\mathbf{x}',\mathbf{t})]\rangle \simeq -\hbar\delta^3(\mathbf{x}-\mathbf{x}')$. This is possible if we replace the result this commutation trough the mean value of z component $f^2_k\langle\hat{K}_{zq}(t)\rangle/\alpha \simeq \tilde{\omega}_q/\omega_q$. More than this, in the expression for $|\mathbf{q}_q|$ we mast replaced the frequency ω_q with its renormalized expression $\tilde{\omega}_q$, so that $|\mathbf{q}_q| = \sqrt{\hbar/(2NM\tilde{\omega}_q)}\, f_q/\sqrt{\alpha}$. About the importance of this re-normalization of electron-electron interaction constants we disuse in the section 4. In this situation the divergence $(\nabla, \mathbf{u}(\mathbf{x})) = \partial_x u_x + \partial_y u_y + \partial_z u_z$ together with the time derivative of the displacement $\partial_t\hat{\mathbf{u}}(\mathbf{x},t)$ must be found from the generalized Heisenberg equations for a positive frequency part of this operator

$$\partial_t\hat{\mathbf{u}}^{(+)}(\mathbf{x},t) = \frac{i}{\hbar}[\hat{H}_v, \hat{\mathbf{u}}^{(+)}(\mathbf{x},t)],\ \partial_i u^{(+)}_i(\mathbf{x},t) = \frac{i}{\hbar}[\mathscr{P}_{vi}, \hat{u}^{(+)}_i(\mathbf{x},t)].$$

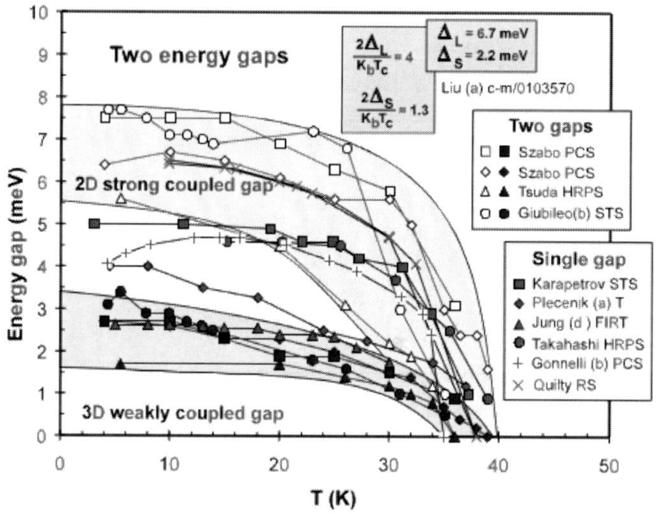

Figure 5. Energy gap dependence on temperature obtained from point contact spectroscopy (PCS), high-resolution photoemission spectroscopy (HRPS), scanning tunnelling spectroscopy (STS), tunnelling (T), far-infra-red transmission (FIRT), Raman spectroscopy (RS) experiments. Data are taken from references [Szabo], [Tsuda], [Giubileo (b)], [Karapetrov], [Plecenik (a)], [Jung (d)], [Takahashi],[Laube], [Gonnelli (b)], [Quilty] (see Ref.[36]).

Using the Hamiltonian (17) and other three components of the energy-momentum tensor $\mathscr{P}_{vi}=\sum_q \pi_{qi}\hat{K}_q^+\,\hat{K}_q^+$ (here $|\pi|=\varepsilon_q/v$) we obtain the following equations for these operators

$$\begin{aligned}\partial_t\hat{\mathbf{u}}^{(+)}(\mathbf{x},t) &\simeq -i\sum_q \tilde{\omega}_q\mathbf{q}_q\hat{K}_q^-(t)\exp(i\mathbf{q}\mathbf{x});\\ \partial_i u_i^{(+)}(\mathbf{x},t) &\simeq -i\sum_q \tilde{q}q_{qi}\hat{K}_q^-(t)\exp(i\mathbf{q}\mathbf{x}).\end{aligned}$$

where the wave vector $\tilde{q}=\tilde{\omega}_q/c$ is normalized like the frequency. Here we have used the following mean value of the operator $\hat{K}_{zq}(t)$. The above description give us the possibility to introduce the normalized frequency $\tilde{\omega}_q$ and momentum $\tilde{q}$ for nonlinear interaction directly in the consultant of interaction of electrons with lattice vibration $Q_1(q)$ 22, which is proportional to the $\sqrt{\tilde{\omega}_q}$.

Taking into consideration the large errors in measurements of the amplitude of order parameter or superconducting gap in $MgCu_2$ [36], we propose to extend our model of intensity coupling vibration modes unifying the decreasing of the re-normalization of the phonon velocities according to the $su(2)$ and increasing of this group velocities with increasing the temperature described by $su(1,1)$ models. Indeed, let us consider that the non-linear parameter κ has the following explicit temperature dependence $\kappa=\kappa_0(T-T_0)/T_c$, where the temperature T_0 is between the 0 and critical values of the temperature phase transition, $0<T_0<T_c$. In this case we observe the decreasing of the re-normalisation of the effective elastic constant (and group velocity of the phonons) till the temperature T_0 and after that its increasing. At first glance, such an explicit temperature dependence is not so clear. As follows from the Review [36] the superconducting gap in the temperature interval $0<T_0<T_c$

defers from one experiment to another. In order to find some correlations between these experiments and theoretical description purposed above is better found this temperature dependence of non-linear from the decomposition principle of the potential energy (7. Indeed, take into consideration the next order in the non-linear decomposition of the potential (4), we may introduce in the Hamiltonian (7) the next contribution

$$\Delta\hat{H} \sim \tilde{B}\hbar\omega_D \sum_{q_1,q_2,q_3} f_{q_1}^2 f_{q_2}^2 f_{q_3}^2 \hat{b}_{q_1}^\dagger \hat{b}_{q_2}^\dagger \hat{b}_{q_3}^\dagger \hat{b}_{q_3} \hat{b}_{q_2} \hat{b}_{q_1}. \tag{24}$$

The single mode along the axis Z may modify the vibration on the layer so that the renormalised operators become: $\hat{K}_{q_\perp}^+ \sim \hat{b}_{q_\perp}^\dagger \sqrt{F(\hat{N}_B)/f}$, ($\hat{K}_{q_\perp}^- = [\hat{K}_{q_\perp}^+]^+$), where $F(N_B) = \alpha - f^2\hat{N}_B + \beta(\hat{N}_B)^2$. In the situation, when, with the increasing of the temperature the number of phonon in the Z mode achieved the value $\alpha = f^2\langle\hat{N}_B\rangle$, the third order (24) may become important in electron phonon interaction $\hat{K}_{q_\perp}^+ \sim \hat{b}_{q_\perp}^\dagger \hat{N}_B\sqrt{\beta}/f$. This effect can be interpreted as a modification of the equilibrium states of the atoms in the layers stimulated by the temperature modification of the structure [37]. In such a lattice representation interpreted by phenomenological operators like Holstain-Primacoff may creates a modification of the amplitude of electron pairing as a function of the displacement of equilibrium states of the nuclei in the temperature interval $0 < T < T_c$, which open the new possibilities in the interpretations of the anomalies in temperature representation of the superconducting observed in the lot of experimental results presented by authors of Ref.[36](see Figure 5).

This effect can be construed as a modification of the equilibrium states of the atoms in the layers with increasing the temperature [37]. In such a lattice representation interpreted by phenomenological operators like to host-Primacoff may creates a modification of the amplitude of electron pairing as a function of the shift of equilibrium states of the nuclei in the temperature interval $0 < T < T_c$, which open the new possibilities in the interpretations of the anomalies in temperature representation of the superconducting, observed in the lot of experimental results presented by authors of Ref.[36](see Figure 5).

The next section 3 is devoted to another nonlinear mechanism described by a two-quantum transitions in multi-band materials. The re-normalization of the interaction constant $Q_1(q)$ is not so significant in the contribution of two-quantum non-linearity in superconductivity.

3. Two-Quantum Transitions between the Electronic States of the Solid

The nonlinear mechanism of electron-phonon interaction may be developed using muti-phonon transitions between the electronic states [6, 19] According to the representation proposed in Sec. 2, we develop below the second model of the deformation of the periodical potential of electrons $V_e(\mathbf{r})$ by the lattice vibration with the value $V_I(r,R) = V_e(\mathbf{r}+\mathbf{u},R) - V_e(r,R)$ proposed in the literature representing the displacement $\mathbf{u}(x)$ through the new cooperative operators $\hat{K}^+$ and $\hat{K}^-$ [18]. Here $V_e(\mathbf{r}+\mathbf{u},R)$ is deformed electronic potential. Decomposing the deformed electronic potential $V_I(r,R)$ on the nonlinear parameter, $\mathbf{u}(x)$: $V_I(r,R) \simeq (\nabla V_e(\mathbf{r},R), \mathbf{u}(\mathbf{x},t))$ and following multi-band representation of electrons in the material, we can introduce the interaction Hamiltonian between the current carriers and nonlinear vibrations of the lattice in the new operator representation

$$\hat{H}^{I}_{e-vib} = -\frac{1}{\sqrt{V}}\sum_{n,m}\sum_{\mathbf{k},\mathbf{q}}\{Q_{m,n}\hat{a}^{\dagger}_{m,k+q}\hat{a}_{n,k}\hat{K}^{-}_{\mathbf{q}} + Q^{*}_{m,n}\hat{a}^{\dagger}_{n,\mathbf{k}}\hat{a}_{m,\mathbf{k+q}}\hat{K}^{\dagger}_{\mathbf{q}}\}, \tag{25}$$

where the interaction constant is represented through the transition matrix element $Q_{m,n} = -\sqrt{\hbar/(2\rho\omega_q)}S_{mn}f_q/\sqrt{\alpha}$, between the n and m bands respectively (see Figure 6)

$$\begin{aligned} S_{mn} &= (\mathbf{e}_q, \int_{\tau_0} d^3\mathbf{r}u^{*}_{m,\mathbf{k+q}}(\mathbf{r})\nabla V_e(\mathbf{r})u_{n,\mathbf{k}}(\mathbf{r})) \\ &= -(\mathbf{e}_q, \int_{\tau_0} d^3\mathbf{r}V_{\mathbf{e}}(\mathbf{r})[u_{\mathbf{n},\mathbf{k}}(\mathbf{r})\nabla u^{*}_{\mathbf{m},\mathbf{k+q}}(\mathbf{r})+u^{*}_{m,\mathbf{k+q}}(\mathbf{r})\nabla u_{n,\mathbf{k}}(\mathbf{r})]). \end{aligned} \tag{26}$$

Here the transition matrix element takes into account the wave vector k spin σ of electrons from the bands n and m , e_q is the polarization vector of vibration mode; $\varepsilon_m(k)$ is the energy of the electron, which can be calculated from the level of the chemical potential; $u_{n,\mathbf{k}}(r)$ is the Bloch amplitude of the wave function. The operators obey the next commutation relations:

$$\left[\hat{a}_{m,k};\hat{a}^{\dagger}_{m_1,k_1}\right]_{+} = \delta_{m,m_1}\delta_{k,k_1}\delta_{\sigma,\sigma_1};\ [\hat{a}_{m,k};\hat{a}_{m_1,k_1}]_{+} = 0, \tag{27}$$

where $a^{\dagger}_{n,k}$ ($(a_{n,k})$) is the electron creation (annihilation) operator in m- band $(m = 1,2)$ with the quasi-wave vector k and spin. The sign $+$ indexes indicate the anti-commutation operations in expressions (27).

The matrix elements of the interaction of the electrons with lattice vibration are known [18]. Therefore the sums on k in the Hamiltonian (25) take into consideration the wave-vector and the spin of electrons in the bands: $k = (k,\sigma)$. Representing the wave function through periodical Bloch amplitude $\psi_{n\kappa}(r) = u_{n,\mathbf{k}}(r)\exp[i(k,r) - i\varepsilon_n(k)t/\hbar]/\sqrt{V}$, we observe that the action of the Hamiltonian of electron in the periodical field $\hat{H}_0 = -\hbar^2\Delta/(2m_e) + V_e(\mathbf{r})$ on these functions give us the following Schrodinger equations for amplitudes $u^{*}_{m,\mathbf{k}_1}$ and $u_{m,\mathbf{k}}$

$$\begin{aligned} \frac{\hbar^2}{2m_e}\Delta u^{*}_{m,\mathbf{k}_1}(\mathbf{r}) \quad & +2i(\mathbf{k}_1, \nabla u^{*}_{m,\mathbf{k}_1}(\mathbf{r})) \\ - \quad & k_1^2 u^{*}_{m,\mathbf{k}_1}(\mathbf{r})\} + (\varepsilon_m(k_1) - V_e(\mathbf{r}))u^{*}_{m,\mathbf{k}_1}(\mathbf{r}) = 0, \end{aligned}$$

$$\frac{\hbar^2}{2m_e}\{\Delta u_{m,\mathbf{k}}(\mathbf{r}) - 2i(\mathbf{k}, \nabla u_{n,\mathbf{k}}(\mathbf{r})) - k^2 u_{n,\mathbf{k}}(\mathbf{r})\} + (\varepsilon_n(k) - V_e(\mathbf{r}))u_{n,\mathbf{k}}(\mathbf{r}) \quad = \quad 0.$$

Here $k_1 = k + q$, $\varepsilon_n(k)$ is the energy of electron in the n-band. Multiplying the first equation with $(\mathbf{e}_q, \nabla u_{n,\mathbf{k}}(r))$ and the second equation with $(e_q, \nabla u^{*}_{m,\mathbf{k}_1}(r))$ and integrating the sum of this equation on the elementary cell of the crystal, one can find the following expression for matrix element (26)

$$\begin{aligned} S_{mn} &= -\frac{i\hbar^2}{m_e}\int_{\tau_0} d^3\mathbf{r}\{(\mathbf{k}+\mathbf{q}, \nabla\mathbf{u}^{*}_{\mathbf{m},\mathbf{k}_{+q}})(\mathbf{e}_\mathbf{q}, \nabla\mathbf{u}_{\mathbf{n},\mathbf{k}}) - (\mathbf{k}, \nabla\mathbf{u}_{\mathbf{n},\mathbf{k}})(\mathbf{e}_\mathbf{q}, \nabla\mathbf{u}^{*}_{\mathbf{m},\mathbf{k+q}})\} \\ & -\mathbf{e}_q\int_{\tau_0} d^3\mathbf{r}\mathbf{u}^{*}_{\mathbf{m},\mathbf{k+q}}\{\varepsilon_m(\mathbf{k}+\mathbf{q}) - \varepsilon_n(\mathbf{k}) + \frac{\hbar^2}{2m_e}(\mathbf{k}^2 - (\mathbf{k}+\mathbf{q})^2)\}\nabla u_{n,\mathbf{k}}. \end{aligned} \tag{28}$$

$$Q_{11}(\mathbf{q}) = -i\frac{f_q}{\sqrt{\alpha}}\sqrt{\frac{\hbar}{2\rho\tilde{\omega}_q}}E_{11}q, \tag{29}$$

where $E_{11} \approx \frac{\hbar^2}{m_e}\int_{\tau_0} d^3\mathbf{r}\{(\mathbf{e_q},\nabla\mathbf{u}^*_{1,\mathbf{k}})(\mathbf{e_q},\nabla\mathbf{u}_{1,\mathbf{k}}) \sim \mu$. In a similar way we obtain the interaction constant for inter-band interaction

$$\begin{aligned} Q_{21}(k,q) &= -\sqrt{\frac{\hbar}{2\rho\tilde{\omega}_q}}\frac{f_q}{\sqrt{\alpha}}(\mathbf{e}_q\int_{\tau_0} d^3\mathbf{r}\mathbf{u}^*_{2,\mathbf{k+q}}\{\varepsilon_2(\mathbf{k}+\mathbf{q})-\varepsilon_1(\mathbf{k}) \\ &+ \frac{\hbar^2}{2m_e}(\mathbf{k}^2-(\mathbf{k}+\mathbf{q})^2)\},\nabla u_{1,\mathbf{k}}) \simeq -\frac{f_q}{\sqrt{\alpha}}\sqrt{\frac{1}{2\rho\omega_q\hbar}}\varepsilon_{21}p_{21}. \end{aligned} \tag{30}$$

Here $\varepsilon_{21} = \varepsilon_2(k+q)-\varepsilon_1(k)$ is the energetic distance between bands one and two (see Figure 6), $p_{21} = \hbar(\mathbf{e}_q,\int_{\tau_0} d^3ru^*_{2,\mathbf{k}+q}\nabla u_{1,\mathbf{k}})$.

In the case when the single quantum transition between the first and the second bands $Q_{21}(q)$ is larger than the matrix element of the single quantum transitions in the intrinsic first band $Q_{11}(q)$, the two-quantum transition between the states $|k\rangle$ and $|k'\rangle$ become larger than the single quantum transitions. This becomes possible, when the first and the second bands are formed from the local atomic levels with different parity. For example, if the first band provided by the local atomic level of S- symmetry crystal, and the second band comes from the level of P- symmetry, it is clear that in the case when $k \sim k'$ the intrinsic matrix element $Q_{11}(q)$ is less than the matrix element between the first and the second bands $Q_{21}(q)$. In this case the two-phonon transitions between the states $|k\rangle$ and $|k'\rangle$ of the first band is similar to two-quantum transitions between the forbidden states $|1S\rangle$ and $|2S\rangle$ of Hydrogen-like atoms an ions (see for example [38, 39]). The second band in such two-quantum transitions plays the role of a virtual state $|P\rangle$.

Taking into account the proprieties of the local symmetry of Bloch wave functions for the down and upper bands, it is not difficult to demonstrate that in many cases when the second and the first bands arise from the atomic local levels with different symmetries (for example P and S atomic states) the inter-band matrix element of the interaction of electrons with lattice vibrations Q_{12} is larger than the intra-band matrix elements Q_{11}, Q_{22} $(Q_{12} \gg Q_{11})$.

These new operators take into account the sign of non-linearity in the vibration lattice subsystem and pass in boson operators for big value of α (small nonlinear parameter κ_q): $\hat{K}^+_{\mathbf{q}} \simeq \hat{b}^\dagger_{\mathbf{q}}\sqrt{\alpha}/f_q$; $\hat{K}^-_{\mathbf{q}} \simeq \hat{b}_{\mathbf{q}}\sqrt{\alpha}/f_q$. The matrix elements pass into electron-phonon interaction constants $G_{m,n} = Q_{m,n}\sqrt{\alpha}/f_q$ well known in the literature [18]. In this section the two-quantum cooperative processes between the electrons of the lowest band through the virtual state of the second band are proposed. For this it is necessary firstly to eliminate the operators of the second band from Hamiltonian (25). If the second band is situated at the energetic distance larger than $\kappa_b T$, we can eliminate the electronic operators of this band from interaction Hamilton (25). In order to solve this problem let us consider the operator $A(t)$ which does not depend on the operators of the second band. The Heisenberg equation

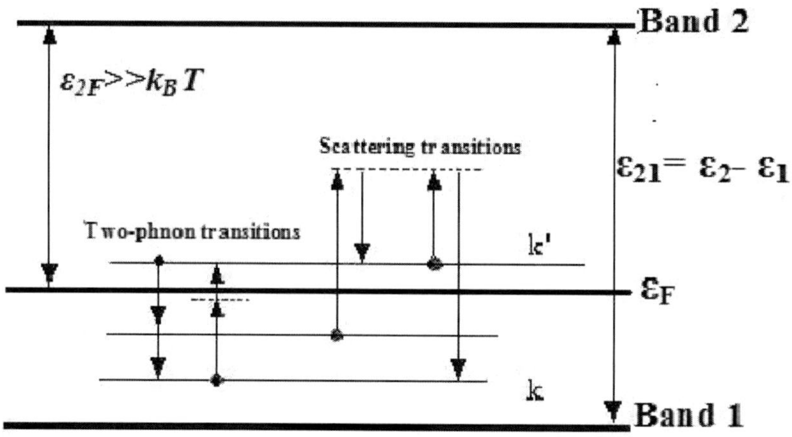

Figure 6. The band structure and two-quantum transitions near the Fermi level. ε_{21} is the energetic distance between the real atomic terms relative, which was constructed the first and second bands structure. $\varepsilon_{2F} = \varepsilon_2(k) - \varepsilon_F$ is the energetic distance between the second band and Fermi level.

for the mean value of this operator $\hat{A}(t)$ is

$$\langle \frac{d\hat{A}(t)}{dt} \rangle = \frac{i}{\hbar} \langle [\hat{H}, \hat{A}(t)] \rangle$$

Using the Hamiltonian (25) one can obtain the following equation for the mean value of operator $\hat{A}(t)$

$$\begin{aligned}
\langle \frac{d\hat{A}(t)}{dt} \rangle &= \frac{i}{\hbar} \sum_k \varepsilon_1(k) \langle [\hat{a}^\dagger_{1k}(t) \hat{a}_{1k}(t), \hat{A}(t)] \rangle \\
&+ \frac{i}{\hbar\sqrt{V}} \sum_{k,\mathbf{q}} \{ Q_{1,1}(\mathbf{q}) \langle [\hat{a}^\dagger_{1,k+\mathbf{q}}(t) \hat{a}_{1,k}(t) \hat{K}^-_{\mathbf{q}}(t), \hat{A}(t)] \rangle \\
&- H.c.t.(\hat{A}^\dagger(t) \to \hat{A}(t)) \} \\
&+ \frac{i}{\hbar\sqrt{V}} \sum_{k,\mathbf{q}} \{ Q_{2,1}(\mathbf{q}) \langle \hat{a}^\dagger_{2,k+\mathbf{q}}(t) [\hat{a}_{1,k}(t) \hat{K}^-_{\mathbf{q}}(t), \hat{A}(t)] \rangle \\
&+ q_{1,2}(\mathbf{q}) \langle [\hat{a}^\dagger_{1,k+\mathbf{q}}(t) \hat{K}^-_{\mathbf{q}}(t), \hat{A}(t)] \hat{a}_{2,k}(t) \rangle \\
&- H.c.t.(\hat{A}^\dagger(t) \to \hat{A}(t)) \}. \qquad (31)
\end{aligned}$$

As the second band plays the role of virtual states for the transitions between the electron states in the first band, it is possible to eliminate the electronic operators of the second band. In this approximation, taking into account the Heisenberg equation for electronic operators from the second band

$$\begin{aligned}
\frac{d\hat{a}_{2,k}(t)}{dt} &= -i\varepsilon_2(k) \hat{a}_{2,k}(t)/\hbar \\
&- \frac{i}{\hbar\sqrt{V}} \sum_{n=1}^{2} \sum_{\mathbf{q}_1} \{ Q_{2n}(\mathbf{q}_1) \hat{a}_{n,k-\mathbf{q}_1}(t) \hat{K}^-_{\mathbf{q}_1}(t) + Q^*_{n2}(\mathbf{q}_1) \hat{K}^\dagger_{\mathbf{q}_1}(t) \hat{a}_{n,k+\mathbf{q}_1}(t) \}
\end{aligned}$$

we obtain the solutions

$$\begin{aligned}
\hat{a}_{2,k}(t) &= \hat{a}_{2,k}(t_0)\exp\left(-\frac{i}{\hbar}\varepsilon_2(k)(t-t_0)\right) \\
&-\frac{i}{\hbar\sqrt{V}}\sum_{n=1}^{2}\sum_{\mathbf{q}_1}\int_{t_0}^{t} d\tau \exp\left(-\frac{i}{\hbar}\varepsilon_2(k)(t-t_0-\tau)\right) \\
&\times\{Q_{2n}(\mathbf{q}_1)\hat{a}_{n,k-\mathbf{q}_1}(\tau)\hat{K}^{-}{}_{\mathbf{q}_1}(\tau)+Q^{*}_{n2}(\mathbf{q}_1)\hat{K}^{\dagger}_{\mathbf{q}_1}(\tau)\hat{a}_{n,k+\mathbf{q}_1}(\tau)\}; \\
\hat{a}^{\dagger}_{2,k}(t) &= [\hat{a}_{2,k}(t)]^{\dagger}. \qquad (32)
\end{aligned}$$

On the right hand site of these operators we have the nonlinear vibration operators $K^{-}{}_{\mathbf{q}_1}(\tau)$ and $K^{\dagger}_{\mathbf{q}_1}(\tau)$. The Heisenberg equation for such operators are

$$\frac{d\hat{K}^{-}{}_{\mathbf{q}}(\tau)}{d\tau} = -2\sigma i\varepsilon_q\hat{K}^{z}{}_{\mathbf{q}}(\tau)\hat{K}^{-}{}_{\mathbf{q}}(\tau)+\sigma\frac{2i}{\hbar\sqrt{V}}\sum_{n,m}\sum_{k}Q^{*}_{m,n}\hat{a}^{\dagger}_{n,\mathbf{k}}(\tau)\hat{K}^{z}_{q}(\tau)\hat{a}_{m,\mathbf{k}+\mathbf{q}}(\tau) \ , \qquad (33)$$

where the signs $\sigma = -1$ and $\sigma = 1$ corresponds to negative and positive values of anharmonic coefficient A respectively. Neglecting the interaction with electronic subsystem we can represent the solution of equation (33) in the following T-product $\hat{K}^{-}{}_{\mathbf{q}}(\tau) = T\exp[\mp 2\sigma i\varepsilon_q\int_0^{\tau}\hat{K}_{zq}(\tau')d\tau']\ \hat{K}^{-}{}_{\mathbf{q}}(0)$. Considering that for short retardation time the operator $\hat{K}^{z}{}_{\mathbf{q}}(\tau')$ can be approximated with the expression its mean value $\langle\hat{K}_{zq}\rangle = \alpha\tilde{\omega}_q/(\omega_q f^2)$, we can represent the solution of operator $\hat{K}^{-}{}_{\mathbf{q}}(\tau)$ in expression (32) thorough the smooth amplitude $\tilde{K}^{-}_{\mathbf{q}}(\tau)$ and rapidly oscillatory part $\exp[\mp 2\sigma i\varepsilon_q\langle\hat{K}^{z}{}_{\mathbf{q}}\rangle\tau]$, so that $\hat{K}^{-}_{\mathbf{q}}(\tau) = \tilde{K}^{-}_{\mathbf{q}}(\tau)\exp[\mp 2\sigma i\varepsilon_q\langle\hat{K}^{z}{}_{\mathbf{q}}\rangle\tau]$. Neglecting the retardation in the smooth amplitude $\tilde{K}^{-}{}_{\mathbf{q}}(\tau)\simeq\tilde{K}^{-}{}_{\mathbf{q}}(t)$ and considering that $|q_{2,2}(q)|\ll|q_{2,1}(q)|$, in the Born-Markoff approximation the operators $a_{2,k}(t)$ and $a^{\dagger}_{2,k}(t)$ can be represented through the vacuum and interaction parts

$$\begin{aligned}
&\hat{a}_{2,k}(t)\simeq\hat{a}_{2,k}(t_0)\exp\left(-\frac{i}{\hbar}\varepsilon_2(k)(t-t_0)\right) \\
&-\frac{1}{\sqrt{V}}\sum_{\mathbf{q}_1}\{\frac{Q_{21}(\mathbf{q}_1)\hat{a}_{1,k-\mathbf{q}_1}\hat{K}^{-}_{\mathbf{q}_1}}{\varepsilon_2(\mathbf{k})-\varepsilon_1(\mathbf{k}-\mathbf{q}_1)-\hbar\tilde{\omega}_{\mathbf{q}_1}} \\
&+\frac{Q^{*}_{12}(\mathbf{q}_1)\hat{K}^{\dagger}_{\mathbf{q}_1}\hat{a}_{1,k+\mathbf{q}_1}}{\varepsilon_2(\mathbf{k})-\varepsilon_1(\mathbf{k}+\mathbf{q}_1)+\hbar\tilde{\omega}_{\mathbf{q}_1}}\}; \\
&\hat{a}^{\dagger}_{2,k}(t) = [\hat{a}_{2,k}(t)]^{\dagger}. \qquad (34)
\end{aligned}$$

Here is taken into account that $\tilde{\omega}_{\mathbf{q}} = 2\sigma\varepsilon_q\langle\hat{K}^{z}{}_{\mathbf{q}}\rangle = \omega_q+\sigma\kappa_q\sum_{q'}f^2_{q'}\langle\hat{n}_{q'}\rangle(1+\delta_{q,q'})$. It is considered that the interaction between the subsystems takes place for time moments $\varepsilon_2 t/\hbar\gg 1$. Taking in to account that the system temperature, T is so less that $\varepsilon_2(k)-\varepsilon_1(k)>>k_bT$ one can consider that the second band remain unpopulated in the processes of two-quantum interactions of electrons from the ground band

$$\lim_{t_0\to-\infty}\hat{a}_{2k}(t_0)|0_2\rangle = 0,\ \lim_{t_0\to-\infty}\langle 0_2|\hat{a}^{\dagger}_{2k}(t_0) = 0,$$

where $|0_2\rangle$ is the vacuum state for the second band. Taking this into consideration, one can eliminate the $a_{2k}(t)$ and $\hat{a}^{\dagger}_{2k}(t)$ operators of the second virtual band after the substitution of the expressions (25) $a_{2,k}(t)$ and $\hat{a}^{\dagger}_{2,k}(t)$ in equation (31).

$$\begin{aligned}
\langle\frac{d\hat{A}(t)}{dt}\rangle &= \frac{i}{\hbar}\langle[\sum_{k}\varepsilon_1(\mathbf{k})\hat{a}^{\dagger}_{1k}\hat{a}_{1k}+\sum_{\mathbf{q}}\hbar\omega_q\hat{K}^{\dagger}_{\mathbf{q}}\hat{K}^{-}_{\mathbf{q}},\hat{A}(t)]\rangle \\
&+ \frac{i}{\hbar\sqrt{V}}\sum_{k,\mathbf{q}}\{Q_{1,1}(\mathbf{q})\langle[\hat{a}^{\dagger}_{1,k+\mathbf{q}}(t)\hat{a}_{1,k}(t)\hat{K}^{-}_{\mathbf{q}}(t),\hat{A}(t)]\rangle \\
&- H.c.t.(\hat{A}^{\dagger}(t)\rightarrow\hat{A}(t))\} \\
&- \frac{i}{\hbar V}\sum_{k}\sum_{\mathbf{q},\mathbf{q}_1}\{Q_{21}(\mathbf{q})Q_{12}(\mathbf{q}_1) \\
&\times \frac{\langle\hat{a}^{\dagger}_{1,k+\mathbf{q}-\mathbf{q}_1}(t)\hat{K}^{\dagger}_{\mathbf{q}_1}(t)[\hat{a}_{1,k}(t)\hat{K}^{-}_{\mathbf{q}}(t),\hat{A}(t)]\rangle}{\varepsilon_2(\mathbf{k}+\mathbf{q})-\varepsilon_1(\mathbf{k}+\mathbf{q}-\mathbf{q}_1)-\hbar\tilde{\omega}_{\mathbf{q}_1}} \\
&+ Q_{21}(\mathbf{q})Q_{12}(\mathbf{q}_1)\frac{\langle\hat{K}^{-}_{\mathbf{q}_1}(t)\hat{a}^{\dagger}_{1,k+\mathbf{q}+\mathbf{q}_1}(t)[\hat{a}_{1,k}(t)\hat{K}^{-}_{\mathbf{q}}(t),\hat{A}(t)]\rangle}{\varepsilon_2(\mathbf{k}+\mathbf{q})-\varepsilon_1(\mathbf{k}+\mathbf{q}+\mathbf{q}_1)+\hbar\tilde{\omega}_{\mathbf{q}_1}} \\
&+ Q_{21}(\mathbf{q})Q^{*}_{21}(\mathbf{q}_1)\frac{\langle[\hat{a}^{\dagger}_{1,k+\mathbf{q}}(t)K^{-}_{\mathbf{q}}(t),\hat{A}(t)]\hat{a}_{1,k-\mathbf{q}_1}(t)K^{-}_{\mathbf{q}_1}(t)\rangle}{\varepsilon_2(\mathbf{k})-\varepsilon_1(\mathbf{k}-\mathbf{q}_1)-\hbar\omega_{\mathbf{q}_1}} \\
&+ Q_{21}(\mathbf{q})Q^{*}_{21}(\mathbf{q}_1)\frac{\langle[\hat{a}^{\dagger}_{1,k+\mathbf{q}}(t)K^{-}_{\mathbf{q}}(t),\hat{A}(t)]K-^{\dagger}_{\mathbf{q}_1}(t)\hat{a}_{1,k+\mathbf{q}_1}(t)\rangle}{\varepsilon_2(\mathbf{k})-\varepsilon_1(\mathbf{k}+\mathbf{q}_1)+\hbar\tilde{\omega}_{\mathbf{q}_1}} \\
&+ H.c.t.(\hat{A}^{\dagger}(t)\rightarrow\hat{A}(t))\}
\end{aligned} \tag{35}$$

The two-quantum exchange between the electrons through the virtual bands is similar to many photon effects in quantum optics and in our opinion has a more realistic diagram interpretation. In analogy with atomic system we can reduce the expression (35) to the Heisenberg equation with Hamiltonian, which takes into consideration two-phonon interaction through the states of virtual band

$$\hat{H} = \hat{H}_0 + \hat{H}_I; \tag{36}$$

where

$$\hat{H}_0 = \sum_{k}(\varepsilon_1(k)-\mu)\hat{a}^{\dagger}_{1k}\hat{a}_{1k}+\sum_{\mathbf{q}}\varepsilon_q\hat{K}^{\dagger}_{\mathbf{q}}\hat{K}^{-}_{\mathbf{q}}$$

is the free part of Hamiltonian for the electronic subsystem in the lower band 1 and lattice vibration subsystem respectively;

$$\hat{H}_I = \hat{H}_{I1} + \hat{H}_{I2}, \quad \hat{H}_{I2} = \hat{H}^{b}_{I2} + \hat{H}^{2s}_{I2} \tag{37}$$

is the interaction Hamiltonian between electron and phonon subsystems which takes in to consideration single quantum transitions

$$\hat{H}_{I1} = \frac{1}{\sqrt{V}}\sum_{k,\mathbf{q}}\{Q_1(\mathbf{q})\hat{a}^{\dagger}_{1k+\mathbf{q}}\hat{a}_{1k}\hat{K}^{-}_{\mathbf{q}} + H.c.\}. \tag{38}$$

The two-quantum transitions and the scattering effects between two quasi-energetic states of the lower band are described by interaction Hamiltonian parts respectively

$$\hat{H}^{b}_{I2} = -\frac{1}{V}\sum_{k}\sum_{\mathbf{q},\mathbf{q}_1}[Q^{b}_{2}(k,q,q_1)\hat{a}^{\dagger}_{1k+\mathbf{q}+\mathbf{q}_1}\hat{a}_{1k}\hat{K}^{-}_{\mathbf{q}_1}\hat{K}^{-}_{\mathbf{q}} + H.c.]; \tag{39}$$

$$\hat{H}^{s}_{I2} = -\frac{1}{V}\sum_{k}\sum_{\mathbf{q},\mathbf{q}_1}[Q^{s}_{2}(k,q,q_1)\hat{a}^{\dagger}_{1k+\mathbf{q}-\mathbf{q}_1}\hat{a}_{1k}\hat{K}^{\dagger}_{\mathbf{q}_1}\hat{K}^{-}_{\mathbf{q}} + H.c.]; \tag{40}$$

Here $a^{\dagger}_{1k}$ and a_{1k} are the creation and annihilation operators of electrons in the lower band of two-band proposed model; $K^{\dagger}_{\mathbf{q}}$ ($K_{\mathbf{q}}$) are creation (annihilation) operator of new renormalized quanta. The interaction constant in single and two-quantum interaction with lattice vibration are $Q_1(\mathbf{q}) = Q_{11}(\mathbf{q})$;

$$Q^{b}_{2}(k,q_1,q_2) = \frac{Q_{21}(\mathbf{q})Q_{12}(\mathbf{q}_1)}{\varepsilon_2(\mathbf{k}+\mathbf{q}) - \varepsilon_1(\mathbf{k}+\mathbf{q}+\mathbf{q}_1) + \hbar\tilde{\omega}_{\mathbf{q}_1}} + \frac{Q_{21}(\mathbf{q})Q_{21}(\mathbf{q}_1)}{\varepsilon_2(\mathbf{k}+\mathbf{q}_1) - \varepsilon_1(\mathbf{k}) - \hbar\tilde{\omega}_{\mathbf{q}_1}}, \tag{41}$$

$$Q^{s}_{2}(k,q_1,q_2) = \frac{Q_{21}(\mathbf{q})Q_{12}(\mathbf{q}_1)}{\varepsilon_2(\mathbf{k}+\mathbf{q}) - \varepsilon_1(\mathbf{k}+\mathbf{q}-\mathbf{q}_1) - \hbar\tilde{\omega}_{\mathbf{q}_1}} + \frac{Q_{21}(\mathbf{q})Q_{21}(\mathbf{q}_1)}{\varepsilon_2(\mathbf{k}-\mathbf{q}_1) - \varepsilon_1(\mathbf{k}) + \hbar\tilde{\omega}_{\mathbf{q}_1}}. \tag{42}$$

We emphasize that free Hamiltonian part contains only the electron states in the lower band and lattice vibration subsystems respectively. The operators $\hat{a}^{\dagger}_{1k}$ and $\hat{a}_{1k}$ are the creation and annihilation operators of electrons in the lower band of two-band proposed model. The energetic distance between the lower and the upper energetically bands satisfies the inequality $\varepsilon_{21} = \varepsilon_2 - \varepsilon_1 > k_b T$. The interaction of electrons in the lower band with the lattice vibrations is described by two types of interaction Hamiltonian $\hat{H}^{(1)}_{I}$ and $\hat{H}^{(2)}_{I}$. In order to neglect the one quantum interaction in comparison with two-quanta we must consider that interaction part $\hat{H}^{(2)}_{I}$ is more large than $\hat{H}^{(1)}_{I}$, or

$$\sum_{q_1}\frac{1}{\sqrt{V}}\left|\frac{Q_{21}(\mathbf{q})Q^{*}_{21}(\mathbf{q}_1)}{\varepsilon_2(\mathbf{k}+\mathbf{q}) - \varepsilon_1(\mathbf{k}+\mathbf{q}\pm\mathbf{q}_1) \pm \hbar\tilde{\omega}_{\mathbf{q}_1}}\right| \gg |Q_{1,1}(\mathbf{q})|. \tag{43}$$

In the next section 4 the single and two-quantum correlation Hamiltonian between the electrons which describe the new type of formation of Cooper pairs it is obtained, taking in consideration the inequality (43) the method of elimination of bi-quantum operators of lattice subsystem is proposed.

4. Correlation Between Single and Two-Phonon Exchange Integrals in Electronic Cooperative Effects

The method in which the single- and two-quanta exchange integrals between the electrons have the same magnitude, $Q_1(q)/(\hbar\sqrt{V}) \sim Q_2(k,q,q_1)/(\hbar V)$, becomes attractive from the physical point of view. In this section we propose the projection operator approach[40] to obtain the master equation of electronic subsystem in single- and two-quantum interaction with lattice vibration [41], which takes into consideration not only the new nonlinear vibration operators, but the two- quantum exchanges between the electrons too. The elimination of the lattice vibration operators, $\hat{K}^{-}_{\mathbf{q}}(t)$ and $\hat{K}^{\dagger}{}_{\mathbf{q}}(t)$, of the vibration subsystem in the master equation was already studying in Ref. [35]. A similar method is proposed in [19] in the opposite approximation $Q_1(q)/(\hbar\sqrt{V}) \ll Q_2(k,q,q_1)/(\hbar V)$. In this section we take into consideration both the mechanisms of non-linearity described above. In order to estimate both mechanisms of non-linearity, we propose to take into consideration the lattice non-linear vibrations together with single and two-quantum transitions. In this case it is necessary to eliminate from the master equation equation not only the $\hat{K}$ operators from the single quantum interaction term, but the two-vibration operator combinations $\hat{K}^{+}_{q}\hat{K}'^{-}_{q}$ of the lattice subsystem, too. According to the Hamiltonian (36) we proposed the equations for the density matrix of the total electron-lattice system in the interaction picture

$$i\hbar\frac{\partial\rho(t)}{\partial t} = [\check{H}_I(t),\rho(t)],$$

where

$$\check{H}_I(t) = \exp\{\frac{i}{\hbar}\hat{H}_o t\}\hat{H}_I\exp\{-\frac{i}{\hbar}\hat{H}_o t\}.$$

Let $\mathscr{P}$ be the projection operator for the complete density matrix $\rho(t)$ on the vector basis of a free vibronic subsystem $\rho_s(t) = \mathscr{P}\rho(t)$, and $\rho_b(t) = \overline{\mathscr{P}}\rho(t)$, where $\rho_s(t)$ and $\rho_b(t)$ are slower and rapidly oscillating parts of density matrix respectively, $\overline{\mathscr{P}} = 1-\mathscr{P}$. It can be shown, that $\mathscr{P}^2 = \mathscr{P}$, $\mathscr{P}\overline{\mathscr{P}} = 0$. Recognizing that for $t = 0$ an electronic subsystem does not interact with lattice vibration, we define the projection operator

$$\mathscr{P} = \rho_{vib}\otimes Tr_{vib}\{\cdots\}, \tag{44}$$

where the trace is taken over the lattice vibration states $|m_q, j_q\rangle$, and $\rho_0 = \rho_e(0)\otimes\rho_{vib}(0)$. Here $\rho_e(0)$ and $\rho_{vib}(0)$ are the density matrix of the free electrons and vibration subsystems, respectively. In this case one can define the slow part of density matrix as, $\check{\rho}_s(t) = \check{\rho}_0\otimes\check{W}(t)$, where $\check{W}(t) = Tr_{vib}\{\rho(t)\}$ is the density matrix of the electronic subsystem. The nonlinear lattice vibration is described now by density matrix $\check{\rho}_{vib}(0) = C\exp[-\beta\sum_q\varepsilon_q\check{K}^{+}_{q}\ \check{K}^{+}_{q}]$, where $\beta = k_BT$, C is the normalization constant. The equations for the matrix $\check{\rho}_s(t)$ and $\check{\rho}_b(t)$ are

$$\frac{\partial\check{\rho}_s(t)}{\partial t} = -i\lambda\mathscr{P}\check{L}_i(t)\{\check{\rho}_s(t)+\check{\rho}_b(t)\}, \tag{45}$$

$$\frac{\partial\check{\rho}_b(t)}{\partial t} = -i\lambda\mathscr{P}\check{L}_i(t)\{\check{\rho}_s(t)+\check{\rho}_b(t)\}, \tag{46}$$

where $\lambda\check{L}_i(t) = [\check{H}_i(t),......]/\hbar$. It is observed that in our definition the projection condition $\mathscr{P}H_I\mathscr{P} = 0$ is not satisfied due to the fact that

$$\mathscr{P}\check{H}^s_{I2}\mathscr{P} = -\frac{1}{V}\sum_k\check{a}^{\dagger}_{1k}\check{a}_{1k}\sum_{\mathbf{q},}Q^s_2(\mathbf{k},\mathbf{q},\mathbf{q})M_q\mathscr{P}$$

Taking this into account, we re-normalized the free energy of electron and lattice vibration subsystems with the expression

$$\check{H}_0 = \sum_k (\varepsilon_1(\mathbf{k}) - \mu)\check{a}^{\dagger}_{1k}\check{a}_{1k} + \sum_q \varepsilon_q \check{K}^{+}_q \check{K}^{+}_q,$$

In this approximation below, we may redefine the two-quantum interaction of electrons with the nonlinear bath of lattice vibration introducing the new interaction Hamiltonian of two-phonon scattering processes

$$\check{H}^s_{I2} = -\frac{1}{V}\sum_k \sum_{\mathbf{q},\mathbf{q}_1} Q^s_2(\mathbf{k},\mathbf{q},\mathbf{q}_1)\check{a}^{\dagger}_{1k+\mathbf{q}-\mathbf{q}_1}\check{a}_{1k}(\check{K}^{\dagger}_{\mathbf{q}_1}\check{K}_{\mathbf{q}} - M_q\delta_{q,q'}) + H.c,$$

So the new Hamiltonian form

$$\check{H}_I(t) = \check{H}_{I1}(t) + \check{H}^{2b}_I(t) + \check{H}^{2s}_I(t)$$

satisfies the projection operator condition $\mathscr{P}\check{H}_I\mathscr{P} = 0$. Here $M_q = \left\langle \check{K}^{\dagger}_{\mathbf{q}_1}\check{K}_{\mathbf{q}} \right\rangle$. Using the distribution function we can estimate this expression

$$\begin{aligned} M_q &= \langle \check{K}^{\dagger}_{\mathbf{q}}\check{K}^{-}_{\mathbf{q}} \rangle = CSp\{\exp[-\check{H}_{vib}/(k_BT)]\check{K}^{\dagger}_{\mathbf{q}}\check{K}^{-}_{\mathbf{q}}\} \\ &= C\sum_{m=0}^{4j(1-\sigma)} [2j_q - \sigma(m-1)]m\exp[-\varepsilon_q[2j_q - \sigma(m-1)]m/(k_BT)]. \end{aligned}$$

This expression tends to the mean number of phonons, N_q, when the parameter α increase $N_q\alpha/f^2_q$, where $N_q = 1/[\exp(\hbar\beta\omega_q) - 1]$. Following the known procedure of elimination of the rapidly oscillating part of the density matrix, we integrate equation(46) relative to the rapid oscillatory part of the density matrix to $\rho_b(t)$. After the substitution of this solution in equation (45), we obtain the expression for the slow part of density matrix

$$i\hbar\frac{\partial\check{\rho}_s(t)}{\partial t} = -\lambda^2 P\int_0^t d\tau \check{L}_i(t)\check{U}(t,t-\tau)\check{L}_i(t-\tau)\check{\rho}_s(t-\tau), \tag{47}$$

where

$$\check{U}(t,t-\tau) = T\exp\left\{-i\lambda\bar{\mathscr{P}}\int_{t-\tau}^{t} d\tau_1\check{L}_i(\tau)\right\}.$$

The quantum correlation between the single and two-quantum interactions of electrons with the lattice vibrations can be found in the third order of the expansion over the small parameter, λ, on the right hand side of equation (47). Confining ourselves to the second and third order of the expansion over the small parameter, λ, we represent the evolution operators $\check{U}(t,t-\tau)$ and $\check{\rho}_s(t-\tau)$ in the following approximate form

$$\check{U}(t,t-\tau) \approx 1 - i\lambda\bar{\mathscr{P}}\int_{t-\tau}^{t} d\tau_1\check{L}_i(\tau_1), \tag{48}$$

and

$$\check{\rho}_s(t-\tau)=\check{\rho}_s(t)+\lambda^2\mathscr{P}\int_0^{\tau}d\tau_1\check{L}_i(t-\tau_1)\int_0^{t-\tau_1}d\tau_2\check{L}_i(t-\tau_1-\tau_2)\check{\rho}_s(t-\tau_1-\tau_2). \quad (49)$$

Upon substitution of equations (48) and 49) in equation (47), in the third order of the small parameter, λ, the equation for $\rho_s(t)$ becomes

$$\begin{aligned}\frac{\partial}{\partial t}\check{\rho}_s(t) &= -\lambda^2\mathscr{P}\int_0^t d\tau\check{L}_i(t)\check{L}_i(t-\tau)\check{\rho}_s(t)\\ &+ i\lambda^3\int_0^t d\tau_1\int_{t-\tau_1}^t d\tau_2\mathscr{P}\check{L}_i(t)\check{L}_i(\tau_2)\check{L}_i(t-\tau_1)\check{\rho}_s(t). \end{aligned}\quad (50)$$

Indeed, considering that $\check{L}_i(t)=[\check{H}_{I1}(t),...]/\hbar+[\check{H}_{I2}(t),...]/\hbar$, and tracing the first term of the expression (50) on the vibration states, we obtain the similar expression as in the last section for single and two-phonon interactions of electron subsystem with thermostat. In this situation, taking into account that the trace of an odd number of $K^{\pm}$ lattice operators takes the zero value, $Tr_{ph}\{\rho_0\check{K}^{\dagger}_{q_1}\check{K}^{-}_{q_2}\check{K}^{\dagger}_{q_3}\}=0$, it is not difficult to observe that the projection of this operator combinations gives the zero contribution in the master equation, $\mathscr{P}\check{H}_{I1}\check{H}_{I2}\mathscr{P}=0$. Following the procedure of calculation of the mean value of $\hat{K}^{\pm}$ operators in analogy with Wick-Bloch-Dominicis theorem [3] it is observed, that only the expressions $\mathscr{P}\check{H}_{I1}\check{H}_{I1}\mathscr{P}\neq 0$ and $\mathscr{P}\check{H}_{I2}\check{H}_{I2}\mathscr{P}\neq 0$ give the main contribution in the master equation

$$\begin{aligned}Tr_{ph}\{\rho_{ph}\check{K}^{+}_{q}\check{K}^{\dagger}_{q_1}\} &= (M_q+\sigma\langle\check{K}_z\rangle)\delta_{q_1,q};\\ Tr_{ph}\{\rho_{ph}\check{K}^{\dagger}_{q_1}\check{K}^{-}_{q}\} &= M_q\delta_{q_1,q};\\ Tr\{\rho_{ph}\check{K}^{-}_{q_1}\check{K}^{-}_{q_3}\check{K}^{\dagger}_{q_2}\check{K}^{\dagger}_{q_4}\} &= (M_{q_1}+\sigma\langle\check{K}_{zq}\rangle)(M_{q_3}+\sigma\langle\check{K}_{zq}\rangle);\\ &\times[\delta_{q_1,q_2}\delta_{q_3,q_4}+\delta_{q_1,q_4}\delta_{q_3,q_2}], \qquad q_2\neq q_4;\\ Tr\{\rho_{ph}\check{K}^{\dagger}_{q_2}\check{K}^{\dagger}_{q_4}\check{K}_{q_1}\check{K}_{q_3}\} &= M_{q_1}M_{q_3}[\delta_{q_1,q_2}\delta_{q_3,q_4}+\delta_{q_1,q_4}\delta_{q_3,q_2}];\quad q_2\neq q_4.\end{aligned}$$

Here

$$\langle\check{K}_{zq}\rangle = \sigma\frac{\alpha}{2f_q^2}+\frac{1}{2f_q^2}\sum_{q'}f_{q'}^2N_{q'}(1+\delta_{q,q'}).$$

Now the single-quantum interaction Hamiltonian can be represented through the annihilation $K^{-}_{\mathbf{q}}$ and creation $\check{K}^{+}_{\mathbf{q}}$ operators of lattice vibration $\check{H}_{I1}(t)=\check{H}^{-}_{I1}(t)+\check{H}^{+}_{I1}(t)$ respectively, so that

$$\check{H}^{-}_{I1}(t)=\frac{1}{\sqrt{V}}\sum_{k,\mathbf{q}}Q_1(\mathbf{q})\check{a}^{\dagger}_{1k+\mathbf{q}}\check{a}_{1k}\check{K}^{-}_{\mathbf{q}},\ \ \check{H}^{+}_{I1}(t)=[\check{H}^{-}_{I1}(t)]^{+}. \quad (51)$$

Following the similar representation one can divide the two-quantum interaction parts (39) and (40) of the Hamiltonian (38), $\check{H}_{I2}(\tau)=\check{H}^{b-}_{I2}(\tau)+\check{H}^{s-}_{I2}(\tau)+\check{H}^{b+}_{I2}(\tau)+\check{H}^{s+}_{I2}(\tau)$, representing two-photon interaction Hamiltonian (39) in the form

$$\begin{aligned}\check{H}^{b-}_{I2} &= -\frac{1}{V}\sum_{k}\sum_{\mathbf{q},\mathbf{q}_1}G^b_2(k,\mathbf{q},\mathbf{q}_1)\check{a}^{\dagger}_{1k+\mathbf{q}+\mathbf{q}_1}\check{a}_{1k}\check{K}_{\mathbf{q}_1}\check{K}_{\mathbf{q}},;\\ \check{H}^{s-}_{I2} &= -\frac{1}{V}\sum_{k}\sum_{\mathbf{q},\mathbf{q}_1}Q^s_2(\mathbf{k},\mathbf{q},\mathbf{q}_1)\check{a}^{\dagger}_{1k+\mathbf{q}-\mathbf{q}_1}\check{a}_{1k}(\check{K}^{\dagger}_{\mathbf{q}_1}\check{K}_{\mathbf{q}}-M_q\delta_{q,q'}),\\ \check{H}^{b+}_{I2} &= [\check{H}^{b-}_{I2}]^{+}\ ,\quad \check{H}^{s+}_{I2}=[\check{H}^{s-}_{I2}]^{+}. \end{aligned}\quad (52)$$

According to this procedure, described by (52), we can obtain from the first term of the right hand site of the equation (50) the following expression

$$\begin{aligned}\Delta\rho_1 &= -\frac{1}{\hbar^2}\int_0^t d\tau\Big\{\Big[\mathscr{P}\check{H}^+_{I1}(t)\check{H}^-_{I1}(t-\tau)\rho_s(t)+\mathscr{P}\check{H}^-_{I1}(t)\check{H}^+_{I1}(t-\tau)\rho_s(t)\Big]\\ &- \Big[\mathscr{P}\check{H}^+_{I1}(t)\rho_s(t)\check{H}^-_{I1}(t-\tau)+\mathscr{P}\check{H}^-_{I1}(t)\rho_s(t)\check{H}^+_{I1}(t-\tau)\Big]\Big\}+H.c., \end{aligned} \quad (53)$$

which contains the single quantum transition between the electrons. In the similar way one can represent the two-quantum exchanges between the electrons of the first band

$$\begin{aligned}\Delta\rho_2 &= -\frac{1}{\hbar^2}\int_0^t d\tau\Big\{\Big[\mathscr{P}\check{H}^{s+}_{I2}(t)\check{H}^{s-}_{I2}(t-\tau)\rho_s(t)+\mathscr{P}\check{H}^{s-}_{I2}(t)\check{H}^{s+}_{I2}(t-\tau)\rho_s(t)\Big]\\ &- \Big[\mathscr{P}\check{H}^{s+}_{I2}(t)\rho(t)\check{H}^{s-}_{I2}(t-\tau)+\mathscr{P}\check{H}^{s-}_{I2}(t)\rho_s(t)\check{H}^{s+}_{I2}(t-\tau)\Big]\\ &+ \Big[\mathscr{P}\check{H}^{b+}_{I2}(t)\check{H}^{b-}_{I2}(t-\tau)\rho_s(t)+\mathscr{P}\check{H}^{b-}_{I2}(t)\check{H}^{b+}_{I2}(t-\tau)\rho_s(t)\Big]\\ &- \Big[\mathscr{P}\check{H}^{b+}_{I2}(t)\rho(t)\check{H}^{b-}_{I2}(t-\tau)+\mathscr{P}\check{H}^{b-}_{I2}(t)\rho_s(t)\check{H}^{b+}_{I2}(t-\tau)\Big]\Big\}+H.c. \end{aligned} \quad (54)$$

This Hamiltonian part represents the two-quantum) cooperative interaction between the electrons, described by the renormalized vibration quanta in sections 2 and 3.

Let us now study the third order contribution in the right hand side of the equation (50). Taking into consideration, that the trace over the odd number of operators gives the zero contribution $\hat{K}_{\pm}$, $Tr_{ph}\{\rho_0\check{K}^{\dagger}_{q_1}\check{K}^{-}_{q_2}\check{K}^{\dagger}_{q_3}\check{K}^{-}_{q_4}\check{K}^{-}_{q_5}\}=0$, it is not difficult to observe, that the projection method over the operator product $\mathscr{P}\check{H}_{I1}\check{H}_{I2}\check{H}_{I2}$ takes the zero value too. In the third order of the small parameter λ the contribution of Liouville operators $\check{L}_{i1}$ and $\check{L}_{i2}$ must be found from the terms like $\mathscr{P}\check{L}_{i1}\check{L}_{i2}\check{L}_{i1}\check{\rho}_s(t)$. For example, the two-quantum contribution in the electron correlations is represented by the following terms

$$\begin{aligned}\Delta\check{\rho}^b_3 &= i\lambda^3\int_0^t d\tau_1\int_0^{\tau_1} d\tau_2\Big\{\mathscr{P}\check{L}^-_{I1}(t)\check{L}^-_{I1}(t-\tau_2)\check{L}^{b+}_{I2}(t-\tau_1)\check{\rho}_s(t)\\ &+ \mathscr{P}\check{L}^-_{I1}(t)\check{L}^{b+}_{I2}(t-\tau_2)\check{L}^-_{I1}(t-\tau_1)\check{\rho}_s(t)\\ &+ \mathscr{P}\check{L}^{b+}_{I2}(t)\check{L}^-_{I1}(t-\tau_2)\check{L}^-_{I1}(t-\tau_1)\check{\rho}_s(t)\Big\}+H.c., \end{aligned} \quad (55)$$

where $\check{L}^-_i(t)=[\check{H}^-_i,...]$. The estimation shows that the main contribution in the correlations between single and two photon interaction with thermostat can be obtained in the case when the two-photon interaction part of the Hamiltonian (52) is represented in the following approximation

$$\begin{aligned}\check{H}^{b-}_{I2}(\tau) &\approx -n\sum_{\mathbf{q}}Q^b_2(\mathbf{q},-\mathbf{q})\check{K}^-_{\mathbf{q}}(\tau)\check{K}^-_{-\mathbf{q}}(\tau),\\ \check{H}^{b+}_{I2}(\tau) &\approx -n\sum_{\mathbf{q}}Q^{*b}_2(\mathbf{q},-\mathbf{q})\check{K}^{\dagger}_{\mathbf{q}}(\tau)\check{K}^{\dagger}_{-\mathbf{q}}(\tau).\end{aligned}$$

Here $n = \sum_k n_k/V$ is the concentration of electrons; $n_k = \langle \breve{a}_k^\dagger \breve{a}_k \rangle = Tr\{\rho_e \breve{a}_k^\dagger \breve{a}_k\}$ is the number of electrons in k state. Representing the Hamiltonian of the single quantum interaction in the form (51) and introducing it together with expression (52) in the equation (55), we obtain the following result

$$\begin{aligned}
\Delta\breve{\rho}_3^b &= -i\frac{1}{\hbar^3}\int_0^t d\tau_1 \int_{t-\tau_1}^t d\tau_2 \{[\mathscr{P}\breve{H}_{I1}^-(t)\breve{H}_{I2}^{b+}(\tau_2)\breve{H}_{I1}^-(t-\tau_1)\breve{\rho}_s(t) \\
&- \mathscr{P}\breve{H}_{I1}^-(t)\breve{H}_{I2}^{b+}(\tau_2)\breve{\rho}_s(t)\breve{H}_{I1}^-(t-\tau_1) - \mathscr{P}\breve{H}_{I1}^-(t)\breve{H}_{I1}^-(t-\tau_1)\breve{\rho}_s(t)\breve{H}_{I2}^{b+}(\tau_2) \\
&+ \mathscr{P}\breve{H}_{I1}^-(t)\breve{\rho}_s(t)\breve{H}_{I1}^-(t-\tau_1)\breve{H}_{I2}^{b+}(\tau_2)] + [\mathscr{P}\breve{H}_{I1}^+(t)\breve{H}_{I2}^{b-}(\tau_2)\breve{H}_{I1}^+(t-\tau_1)\breve{\rho}_s(t) \\
&- \mathscr{P}\breve{H}_{I1}^+(t)\breve{H}_{I2}^{b-}(\tau_2)\breve{\rho}_s(t)\breve{H}_{I1}^+(t-\tau_1) + \mathscr{P}\breve{H}_{I1}^+(t)\breve{\rho}_s(t)\breve{H}_{I1}^+(t-\tau_1)\breve{H}_{I2}^{b-}(\tau_2) \\
&- \mathscr{P}\breve{H}_{I1}^+(t)\breve{H}_{I1}^+(t-\tau_1)\breve{\rho}_s(t)\breve{H}_{I2}^{b-}(\tau_2)]\} + H.c. \qquad (56)
\end{aligned}$$

In the similar situation one can take into account the resonances in scattering interaction between the electrons represented in Figure 6

$$\begin{aligned}
\Delta\breve{\rho}_3^s = i\lambda^3 \int_0^t d\tau_1 \int_0^{\tau_1} d\tau_2 \Big\{ &\mathscr{P}\breve{L}_{I1}^-(t)\breve{L}_{I1}^+(t-\tau_2)\breve{L}_{I2}^{s+}(t-\tau_1)\breve{\rho}_s(t) \\
&+\mathscr{P}\breve{L}_{I1}^+(t)\breve{L}_{I1}^-(t-\tau_2)\breve{L}_{I2}^{s+}(t-\tau_1)\breve{\rho}_s(t) \\
&+\mathscr{P}\breve{L}_{I1}^-(t)\breve{L}_{I2}^{s+}(t-\tau_1)\breve{L}_{I1}^+(t-\tau_2)\breve{\rho}_s(t) \\
&+\mathscr{P}\breve{L}_{I1}^+(t)\breve{L}_{I2}^{s+}(t-\tau_1)\breve{L}_{I1}^-(t-\tau_2)\breve{\rho}_s(t) \\
&+\mathscr{P}\breve{L}_{I2}^{s+}(t)\breve{L}_{I1}^-(t-\tau_1)\breve{L}_{I1}^+(t-\tau_2)\breve{\rho}_s(t) \\
&+\mathscr{P}\breve{L}_{I2}^{s+}(t)\breve{L}_{I1}^+(t-\tau_1)\breve{L}_{I1}^-(t-\tau_2)\breve{\rho}_s(t)\Big\} + H.c.. \qquad (57)
\end{aligned}$$

According with these expressions (53), (54) and (56), we obtain the following master equation for density matrix of electronic subsystem $\breve{W}(t)$

$$\frac{d\breve{W}(t)}{dt} = \frac{d\breve{W}^{(1)}(t)}{dt} + \frac{d\breve{W}^{(2)}(t)}{dt} + \frac{d\breve{W}^{(12)}(t)}{dt}. \qquad (58)$$

The trace of the right hand site of expressions (53) and (54) give us the following results

$$\begin{aligned}
Tr_{ph}\{\rho_{ph}\breve{K}_q\breve{K}_{q_1}^\dagger\} &= (M_q + \sigma\langle\breve{K}_{zq}\rangle)\delta_{q_1,q};\ Tr_{ph}\{\rho_{ph}\breve{K}_{q_1}^\dagger\breve{K}_q^-\} = M_q\delta_{q_1,q}, \\
Tr\{\rho_{ph}\breve{K}_{q_1}\breve{K}_{q_3}\breve{K}_{q_2}^\dagger\breve{K}_{q_4}^\dagger\} &= (M_{q_1} + +\sigma\langle\breve{K}_{zq}\rangle)(M_{q_3} + \sigma\langle\breve{K}_{zq}\rangle) \\
&\times\ [\delta_{q_1,q_2}\delta_{q_3,q_4} + \delta_{q_1,q_4}\delta_{q_3,q_2}], \qquad q_2 \neq q_4, \\
Tr\{\rho_{ph}\breve{K}_{q_2}^\dagger\breve{K}_{q_4}^\dagger\breve{K}_{q_1}^-\breve{K}_{q_3}^-\} &= M_{q_1}M_{q_3}[\delta_{q_1,q_2}\delta_{q_3,q_4} + \delta_{q_1,q_4}\delta_{q_3,q_2}];\quad q_2 \neq q_4.
\end{aligned}$$

Taking into account that the integral on the retardation τ can be represented through $\zeta(x)$ function

$$\begin{aligned}
&i\zeta(\tilde{\omega}_q - [\varepsilon({}_1\mathbf{k}+\mathbf{q}) - \varepsilon_1(\mathbf{k})]/\hbar) = \\
&\lim_{t\to\infty}\int_0^t d\tau \exp(-i[\tilde{\omega}_q - [\varepsilon_1(\mathbf{k}+\mathbf{q}) - \varepsilon_1(\mathbf{k})]/\hbar - i\varepsilon]\tau) = \\
&\frac{-i}{\tilde{\omega}_q - [\varepsilon_1(\mathbf{k}+\mathbf{q}) - \varepsilon_1(\mathbf{k})]/\hbar - i\varepsilon},
\end{aligned}$$

we obtain the following contributions to right hand side of the master equation, described by the single-quantum interaction (53)

$$\frac{d\breve{W}^{(I)}(t)}{dt} = -\frac{i}{\hbar^2 V}\sum_{k,k_1}\sum_{\mathbf{q}}\frac{|Q_1(\mathbf{q})|^2}{[\tilde{\omega}_q - [\varepsilon(\mathbf{k}+\mathbf{q}) - \varepsilon(\mathbf{k})]/\hbar - i\varepsilon}$$
$$\times \left[M_q\langle[[\breve{a}^\dagger_{k_1}\breve{a}_{k_1+q}(t), \breve{W}(t)], \breve{a}^\dagger_{k+q}\breve{a}_k]\rangle + \sigma\langle K_{zq}\rangle\langle[\breve{a}^\dagger_{k_1}\breve{a}_{k_1+q}, \breve{W}(t)]\breve{a}^\dagger_{k+q}\breve{a}_k\rangle\right]$$
$$+H.c. \tag{59}$$

The two-quantum contribution to exchange process between the electrons is obtained from (54)

$$\begin{aligned}\frac{d\breve{W}^{(2)}(t)}{dt} &= +\frac{i}{\hbar V^2}\sum_{k,k_1}\sum_{\mathbf{q},\mathbf{q}_1} Q_2^s(\mathbf{k},\mathbf{q},\mathbf{q}_1)Q_2^s(\mathbf{k}_1,\mathbf{q}_1,\mathbf{q})\\
&\times\ \zeta^*(\hbar(\omega_{\mathbf{q}_1} - \omega_{\mathbf{q}}) - \varepsilon_1(\mathbf{k}_1+\mathbf{q}_1-\mathbf{q}) + \varepsilon_1(\mathbf{k}_1))\\
&\times\ \{U_q(M_{q_1} + \sigma\langle K_{zq}\rangle)[\breve{a}^\dagger_{k+q-q_1}\breve{a}_{\mathbf{k}}, \breve{W}(t)\breve{a}^\dagger_{k_1+q_1-q}\breve{a}_{\mathbf{k}_1}]\\
&+\ U_{q_1}(M_q + \sigma\langle K_{zq}\rangle)[\breve{a}^\dagger_{\mathbf{k}_1+\mathbf{q}_1-\mathbf{q}}\breve{a}_{\mathbf{k}_1}\breve{W}(t), \breve{a}^\dagger_{k+\mathbf{q}-\mathbf{q}_1}\breve{a}_k]\}\\
&+\ \frac{i}{\hbar V^2}\sum_{k,k_1}\sum_{\mathbf{q},\mathbf{q}_1} Q_2^b(\mathbf{k},\mathbf{q},\mathbf{q}_1)Q_2^b(\mathbf{k}_1,\mathbf{q}_1,\mathbf{q})\{(M_q + \sigma\langle\breve{K}_{zq}\rangle)\\
&\times\ (M_{q_1} + \sigma\langle\breve{K}_{zq}\rangle)[\breve{a}^\dagger_{k_1}\breve{a}_{k_1+q+q_1}\breve{W}(t), \breve{a}^\dagger_{k+q+q_1}\breve{a}_k]\\
&-\ U_qU_{q_1}[\breve{W}(t)\breve{a}^\dagger_{k_1}(t)\breve{a}_{k_1+\mathbf{q}+\mathbf{q}_1}(t), \breve{a}^\dagger_{k+q+q_1}(t)\breve{a}_k(t)]\}\\
&\times\ \zeta^*(\hbar(\omega_{\mathbf{q}_1} + \omega_{\mathbf{q}}) - \varepsilon_1(\mathbf{k}_1+\mathbf{q}_1+\mathbf{q}) + \varepsilon_1(\mathbf{k}_1)) + H.c.\end{aligned} \tag{60}$$

Let us now separate the Hamiltonian parts of the equations (59) and (60). Taking into account the imaginary part of the function $i\zeta(x)$ in master equations (59) and (60), we obtain the following explicit expressions of collective interaction part of the Hamiltonian

$$\breve{H}_I^{eff} = \breve{H}_{I1}^{eff} + \breve{H}_{I2}^{eff}. \tag{61}$$

Here the first term corresponds to BCS interaction Hamiltonian [35] normalized by the nonlinear lattice vibration

$$\breve{H}_{I1}^{eff} = -\frac{1}{V}\sum_{k,k_1}\sum_{\mathbf{q},}\frac{\hbar|G_1(q)|^2\tilde{\omega}_{\mathbf{q}}[1+\sigma\sum_{q'} f_{q'}^2 N_{q'}(1+\delta_{q,q'})/\alpha]}{[\hbar\tilde{\omega}_{\mathbf{q}}]^2 - [\varepsilon_1(\mathbf{k}_1+\mathbf{q}) - \varepsilon_1(\mathbf{k}_1)]^2}\breve{a}^\dagger_{\mathbf{k}+\mathbf{q}}\breve{a}^\dagger_{\mathbf{k}_1}\breve{a}_{\mathbf{k}_1+\mathbf{q}}\breve{a}_{\mathbf{k}}, \tag{62}$$

where instead of the constant $|Q_1(q)|^2 = f_k^2/\alpha|G_1(\tilde{\omega}_q)|^2$ we introduced the well known constant of electron-phonon interaction $|G_1(\tilde{\omega}_q)|^2$ like in the BCS Hamiltonian with re-normalized frequency. Returning to the re-normalization of the phonon frequency $\tilde{\omega}_q = \omega_q[1+\sigma\sum_{q'} f_{q'}^2 N_{q'}(1+\delta_{q,q'})/\alpha]$, we observe that in the elastic case, when $\varepsilon_1(\mathbf{k}_1+\mathbf{q}) - \varepsilon_1(\mathbf{k}_1) \simeq 0$ the temperature dependence of interaction energy, $|G_1(\tilde{\omega}_q)|^2 = \Lambda\hbar\tilde{\omega}_q$, gives us the following temperature dependence of single-quantum interaction between the electrons $|G_1(\tilde{\omega}_q)|^2\tilde{\omega}_q[1+\sigma\sum_{q'} f_{q'}^2 N_{q'}(1+\delta_{q,q'})/\alpha]/\tilde{\omega}_{\mathbf{q}}^2 = \Lambda[1+\sigma\sum_{q'} f_{q'}^2 N_{q'}(1+\delta_{q,q'})/\alpha]$.

The second term corresponds to two-quantum correlations between the carriers is described by the following interaction part of the Hamiltonian

$$\begin{aligned}\check{H}_{12}^{eff} &= -\sum_{k,k_1}\sum_{\mathbf{q},\mathbf{q}_1}\Big[M^b_{\mathbf{q},\mathbf{q}'}(k,k_1)[\sigma\sum_{q'} f^2_{q'}N_{q'}(1+\delta_{q,q'})/\alpha+1+N_q+N_{q_1}] \\ &\quad\times\check{a}^\dagger_{k+q+q_1}(t)\check{a}_{\mathbf{k}}(t)\check{a}^\dagger_{\mathbf{k}_1}(t)\check{a}_{k_1+q+q_1}(t) \\ &+ M^s_{\mathbf{q},\mathbf{q}'}(k,k_1)(N_q-N_{q_1})\check{a}^\dagger_{k_1+q_1-q}(t)\check{a}_{k_1}(t)\check{a}^\dagger_{k+q-q_1}(t)\check{a}_k(t)\Big], \end{aligned} \tag{63}$$

which pass into the two-phonon interaction Hamiltonian studied in the Refs.[19],[43], when the nonlinear parameter takes zero value, $A = 0$. This corresponds to the limit $\alpha \to \infty$. In order to compare with the existing approach, we can pass to the traditional electron-phonon constants $|Q_i(q)|^2 = f_k^2/\alpha|G_i(\tilde{\omega}_q)|^2$ in the two-quantum interaction Hamiltonian part (63). Here the function $M^b_{\mathbf{q},\mathbf{q}'}(k,k_1)(1+N_q+N_{q_1})$ represents the exchange energy between two electrons from the states k, and k' in the two-quantum absorption and emission processes

$$\begin{aligned} M^b_{\mathbf{q},\mathbf{q}'}(k,k_1) &= \frac{1}{2V^2}\Bigg[\frac{G_2^b(k,\mathbf{q},\mathbf{q}_1)G_2^{*b}(k_1,\mathbf{q}_1,\mathbf{q})}{\hbar(\tilde{\omega}_{\mathbf{q}_1}+\tilde{\omega}_{\mathbf{q}})-\varepsilon_1(\mathbf{k}_1+\mathbf{q}_1+\mathbf{q})+\varepsilon_1(\mathbf{k}_1)} \\ &+ \frac{G_2^b(k,\mathbf{q},\mathbf{q}_1)G_2^{*b}(k_1,\mathbf{q}_1,\mathbf{q})}{\hbar(\tilde{\omega}_{\mathbf{q}_1}+\tilde{\omega}_{\mathbf{q}})-\varepsilon_1(\mathbf{k}+\mathbf{q}_1+\mathbf{q})+\varepsilon_1(\mathbf{k})}\Bigg]. \end{aligned}$$

Here $M^s_{\mathbf{q},\mathbf{q}'}(k,k_1)(N_q-N_{q_1})$ is the exchange energy between two electrons from the same state in two-quantum scattering processes

$$M^s_{\mathbf{q},\mathbf{q}'}(k,k_1) = \frac{1}{V^2}\frac{G_2^s(k,\mathbf{q},\mathbf{q}_1)G_2^{*s}(k_1,\mathbf{q}_1,\mathbf{q})}{\hbar(\tilde{\omega}_{\mathbf{q}_1}-\tilde{\omega}_{\mathbf{q}})-\varepsilon_1(\mathbf{k}_1+\mathbf{q}_1-\mathbf{q})+\varepsilon_1(\mathbf{k}_1)}.$$

It is observed here that for elastic scattering process $\varepsilon_1(\mathbf{k}_1+\mathbf{q}_1-\mathbf{q}) \approx \varepsilon_1(k_1)$ and $q \sim 2k_F$ the contribution of scattering terms in superconductivity is positive

$$\frac{N_q-N_{q_1}}{\tilde{\omega}_{\mathbf{q}_1}-\tilde{\omega}_{\mathbf{q}}} > 0, \quad N_{q_1} < N_q \; when \; \tilde{\omega}_{\mathbf{q}_1} > \tilde{\omega}_{\mathbf{q}}.$$

We observe here that both nonlinear processes connected with the transitions through the virtual bands and an-harmonic process of the lattice vibration give a significant contribution in two-quantum transition between the electrons. The contribution of nonlinear lattice vibration increases the cooperative interaction between the electrons through the positive an-harmonic potential.

Now we can return to the expression for interference term $d\check{W}^{(12)}/dt$ of the master equation (58). The contribution of this process can be obtained from the second term of equation,(50), proportional to the λ^3. Taking into account the definition of projection operator $\mathscr{P}$, (44), we obtain the following representation for the trace on the lattice vibration

$$\begin{aligned}
Tr \quad & \{\check{\rho}_{ph}\check{K}^-_{q_1}(t)\check{K}^-_{q_2}(t-\tau_1)\check{K}^\dagger_{\mathbf{q}}(\tau_2)\check{K}^\dagger_{-\mathbf{q}}(\tau_2)\} \\
= \quad & (\delta_{\mathbf{q}_1,\mathbf{q}}\delta_{\mathbf{q}_2,-\mathbf{q}}+\delta_{\mathbf{q}2,\mathbf{q}}\delta_{\mathbf{q}_1,-\mathbf{q}})(M_q+\sigma\langle\check{K}_{zq}\rangle)^2 \\
\times \quad & \exp[-2i\tilde{\omega}_q t+2i\tilde{\omega}_q\tau_2+i\tilde{\omega}_q\tau_1]; \\
Tr \quad & \{\check{\rho}_{ph}\check{K}^-_{q_1}(t)\check{K}^\dagger_{\mathbf{q}}(\tau_2)\check{K}^\dagger_{-\mathbf{q}}(\tau)\check{K}^-_{q_2}(t-\tau_1)\} \\
= \quad & (\delta_{\mathbf{q}_1,\mathbf{q}}\delta_{\mathbf{q}_2,-\mathbf{q}}+\delta_{\mathbf{q}2,\mathbf{q}}\delta_{\mathbf{q}_1,-\mathbf{q}})N_q(N_q+\sigma\langle\check{K}_{zq}\rangle) \\
\times \quad & \exp[-2i\tilde{\omega}_q t+2i\tilde{\omega}_q\tau_2+i\tilde{\omega}_q\tau_1];
\end{aligned} \tag{64}$$

$$\begin{aligned}
Tr \quad & \{\check{\rho}_{vib}\check{K}^\dagger_{q_1}(t)\check{K}^\dagger_{q_2}(t-\tau_1)\check{K}^-_{\mathbf{q}}(\tau_2)\check{K}^-_{-\mathbf{q}}(\tau_2)\} \\
= \quad & (\delta_{\mathbf{q}_1,\mathbf{q}}\delta_{\mathbf{q}_2,-\mathbf{q}}+\delta_{\mathbf{q}2,\mathbf{q}}\delta_{\mathbf{q}_1,-\mathbf{q}})M_q^{\;2} \\
\times \quad & \exp[+2i\tilde{\omega}_q t-2i\tilde{\omega}_q\tau_2-i\tilde{\omega}_q\tau_2]; \\
Tr \quad & \{\check{\rho}_{vib}\check{K}^\dagger_{q_1}(t)\check{K}^-_{\mathbf{q}}(\tau_2)\check{K}^-_{-\mathbf{q}}(\tau_2)\check{K}^\dagger_{q_2}(t-\tau_1)\} \\
= \quad & (\delta_{\mathbf{q}_1,\mathbf{q}}\delta_{\mathbf{q}_2,-\mathbf{q}}+\delta_{\mathbf{q}2,\mathbf{q}}\delta_{\mathbf{q}_1,-\mathbf{q}})M_q(M_q+\sigma\langle\check{K}_{zq}\rangle) \\
\times \quad & \exp[+2i\tilde{\omega}_q t-2i\tilde{\omega}_q\tau_2-i\tilde{\omega}_q\tau_2].
\end{aligned} \tag{65}$$

According to the expression (64), the integration over the variable τ in the first two terms of expression (56) is represented by the relation

$$\begin{aligned}
J_1 &= \exp[-2i\omega_q t]\int_0^t d\tau_1\int_{t-\tau_1}^t d\tau_2\exp[2i\tilde{\omega}_q\tau_2+i[\tilde{\omega}_q-(\varepsilon_{k'+q}-\varepsilon_{k'})/\hbar+i\gamma]\tau_1] \\
&= \frac{1}{2i\omega_q}\int_0^t d\tau_1[1-\exp[-2i\tilde{\omega}_q\tau_1]]\exp[i[\tilde{\omega}_q-(\varepsilon_{k'+q}-\varepsilon_{k'})/\hbar+i\gamma]\tau_1] \\
&= \frac{1}{2i\tilde{\omega}_q}\Big[\frac{\exp\{-i[(\varepsilon_{k'+q}-\varepsilon_{k'})/\hbar-\tilde{\omega}_q-i\gamma]t\}-1}{-i[(\varepsilon_{k'+q}-\varepsilon_{k'})/\hbar-\tilde{\omega}_q-i\gamma]} \\
&\quad -\frac{\exp\{-i[(\varepsilon_{k'+q}-\varepsilon_{k'})/\hbar+\tilde{\omega}_q-i\gamma]t\}-1}{-i[(\varepsilon_{k'+q}-\varepsilon_{k'})/\hbar+\tilde{\omega}_q-i\gamma]}\Big].
\end{aligned}$$

For $\gamma t >> 1$ this expression takes the following approximate value

$$J_1 = -\frac{1}{[(\varepsilon_{k'+q}-\varepsilon_{k'})/\hbar-i\gamma)^2-\tilde{\omega}_q^2]}.$$

The time integrals in the last two terms in the expression (56) give the Hermit conjugate form of the integral J_1^*. In this situation, according to the definition $\Delta\rho_3^{(1)}=\rho_0\times dW^{(12)}/dt$, we obtain the following contribution in the interference part of the right hand site of equation (58)

$$\begin{aligned}
\frac{d\breve{W}^{(12)}}{dt} &= +2i\sum_{k,k'}\sum_{\mathbf{q}} Q_1(\mathbf{q})Q_1(-\mathbf{q})Q_2^{*b}(\mathbf{q},-\mathbf{q})(M_q+\sigma\langle\breve{K}_{zq}\rangle)\\
&\times [\breve{a}^{\dagger}_{k+\mathbf{q}}\breve{a}_{1k}\breve{a}^{\dagger}_{k'-\mathbf{q}}\breve{a}_{k'}\breve{W}(t)-\breve{a}^{\dagger}_{k+\mathbf{q}}\breve{a}_{1k}\breve{W}(t)\breve{a}^{\dagger}_{k'-\mathbf{q}}\breve{a}_{1k'}]\\
&\times \frac{1}{(\varepsilon_1(k'+q)-\varepsilon_1(k'))/\hbar-i\gamma)^2-\tilde{\omega}_q^2}\\
&- 2i\sum_{k,k'}\sum_{\mathbf{q}} Q_1^*(\mathbf{q})Q_1^*(-\mathbf{q})Q_2^b(\mathbf{q},-\mathbf{q})M_q\\
&\times [\breve{a}_{1k}\breve{a}^{\dagger}_{1k+\mathbf{q}}\breve{a}_{1k'}\breve{a}^{\dagger}_{1k'-\mathbf{q}}\breve{W}(t)-\breve{a}_k\breve{a}^{\dagger}_{k+\mathbf{q}}\breve{W}(t)\breve{a}_{k'}\breve{a}^{\dagger}_{k'-\mathbf{q}}]\\
&\times \frac{1}{(\varepsilon_1(k'+q)-\varepsilon(k'))/\hbar+i\gamma)^2-\tilde{\omega}_q^2}+H.c. \qquad (66)
\end{aligned}$$

Finding the commutation expression for imaginary part of third term in from the equation (66), we obtain the interference part of single and two-quantum processes of the interaction Hamiltonia

$$\begin{aligned}
\breve{H}^{eff}_{I12} &= \frac{2n}{V}\sum_{k,k_1}\sum_{\mathbf{q},}\frac{|G_1(\mathbf{q})G_1(-\mathbf{q})G_2^b(\mathbf{q},-\mathbf{q})|\cos[2\varphi_q-\varphi_0]}{(\hbar\tilde{\omega}_q)^2-[\varepsilon_1(k_1+q)-\varepsilon_1(k_1)]^2}\\
&\times[\sigma\sum_{q'}f^2_{q'}N_{q'}(1+\delta_{q,q'})/\alpha+1]\breve{a}^{\dagger}_{k+q}\breve{a}_{k_1}{}^{\dagger}\breve{a}_{k_1+q}\breve{a}_k. \qquad (67)
\end{aligned}$$

The effective interaction Hamiltonian parts (62), (63) and (67) between the electrons from the first band through the single and two-quantum of letter vibrations has been obtained in this section. The cooperative interaction between the electrons through the single and two-quantum field is simplified not only to the single-quantum processes of simultaneous absorption or emission of the virtual vibration quanta described by the new operators (9-18), but includes the two-quantum exchanges between the electrons using such an intensity coupling field representation, too. As a consequence of the effective interaction Hamiltonian (62) describes the transitions between two states of the first band, and takes place with a single K-quantum absorption and the emission of another quasi-particle of lattice vibration field. The non-boson character of the new operator, which re-normalases the traditional BCS exchange integral between Cooper pairs as the function of the sign of nonlinear parameter is given, takes into consideration the single- and two-quantum changes between the electrons.

As follows from the effective Hamiltonian parts (62), (63) and (67), we can neglect the interference in the effective interaction Hamiltonian in the case when $2|nG_2^b(\mathbf{q},-\mathbf{q})|\cos[2\varphi_1-\varphi_b]/\hbar\tilde{\omega}_q << 1$. In this situation the single- and two-quantum exchange integrals are proportional to λ^2 while the interference contribution is neglected. In opposite case $2|nG_2^b(\mathbf{q},-\mathbf{q})|\cos[2\varphi_1-\varphi_b]/\hbar\tilde{\omega}_q \sim 1$ the contribution of interference effect drastically changes the BCS Hamiltonian (62) and the contribution of non-temperature dependent exchange integral in interaction Hamiltonian. This term diminishes the BCS contribution in superconductivity when $\cos[2\varphi_1-\varphi_b]=1$ and $\chi_q>0$.

In the next section we take into consideration the contribution of the lattice an-harmonic vibration to the single- and two-quantum exchanges between the electrons in superconductivity phase. The non-linear cooperative exchanges between the electrons increases with the increasing of temperature if we take into consideration these effects. The effective Hamiltonian depends on temperature through the mean number of quasi-particles in the bath, M_q, described by two mechanisms of non-linearity: an-harmonic vibration of lattice and the two-quantum exchange between the electrons. The re-normalization of BCS attractive potential between the electrons by lattice non-linearity and two-quantum exchanges brings to the increasing of effective Debye frequency. These contributions were described in the language of the new lattice vibration operators which have re-normalized the order parameter as a function of the temperature at the first stage of superconductivity. These effects become small if we neglect the lattice an-harmonic vibration and two-phonon exchanges between the electrons. In opposite case, these non-linear effects may give the substantial contribution in the re-normalization of the exchange energy between the Cooper pairs and the value of the critical temperature in superconductivity. The contribution of the correlations between the single and two-quantum exchanges between the electrons is also proposed for the numerical studies in the section 5.

5. Two Mechanisms for Temperature Modification of Order Parameter in Nonlinear Superconductivity

In order to simplify the numerical results, in this section we will approximate the mean number of lattice vibration quanta M_q with the main phonon number. Following the conception about two possibilities of the modification of the order parameter in superconductivity, we propose the separate discussion of the numerical approach to each model. As the influence of anisotropy on the vibration dispersion law of the vibration modes in the layers of the superconducting materials was regarded in Sec. 2 as a possibility of the re-normalization of limit value of the Debye frequency as the function of temperature in the point *I*., we will take it into consideration. We are interested in the increasing of order parameter discussed in the BCS model, due to the possibilities of the modification of elastic constant of the vibrations in the layers (for example, the *Cu*-atoms in the material $MgCu_2$) stimulated by other vibrational modes. As the critical temperature in BCS model increases with the increasing of the Debye frequency in point *I*., we take into consideration the positive value of the non-linear parameter, κ, described in the Sec.2. The anisotropy of the interaction between vibrations in σ bounds of the layers and vibration perpendicular to the multilayer structure (π -vibrations) in $MgCu2$ opens the possibilities to change the critical temperature using the external sources of the excitations of the vibrations.

Another mechanism of non-linearity in superconducting materials may be the strong coupling of electron phonon modes so that two-phonon transitions may be more effective than traditional single-phonon one. In this position of the second virtual band relative the Fermi level plays an important role in the creation of the two-phonon Cooper pairing effect on the first conducting band. For example, if the second virtual state (bound level or band) is situated near the Fermi level or in other non-conducting layer of the material, the two-phonon scattering effects with absorption of one phonon from the thermostat and generation of another phonon may play a more important role than processes with simultaneous generation or absorption of phonon pairs. We discuss a simpler effect where the distance between the first and second band is larger than the mean value of the photon energy. As a consequence of classical approximations in superconductivity, only the electrons which have opposite momentum and spins are considered and they are situated in a Debye layer near the Fermi surface. In this case the interaction corresponds to an attraction potential between the electrons. In this situation in expressions (62), (63) and (67) it is assumed that $\hbar\omega_{\mathbf{q}} \ll \varepsilon_2(k+q) - \varepsilon_1(k)$ and $1/(\hbar\omega_{\mathbf{q}} - (\varepsilon_2(k+q) - \varepsilon_1(k)) \approx 1/\varepsilon_{21}$. Here the ε_{21} can be approximated with the energy distance between the bottom of second band and Fermi surface, $\varepsilon_{21} \approx \varepsilon_2(0) - \mu$, when the Fermi energy μ is less than ε_{21}. Let us firstly take into account the Hamiltonian parts (62) and (61). Replacing $k_1 \to -k_1$ and considering that the main contribution $q = k_1 - k$ the following effective interaction Hamiltonian between the electrons is obtained from these two parts

$$\check{H}_{Ieff12} = -\frac{1}{V}\sum_{k,k_1} G^{(1)}(k,k_1)\check{a}^{\dagger}_{k_1}\check{a}^{\dagger}_{-k_1}\check{a}_{-k}\check{a}_k. \tag{68}$$

According to the approximated representation, the matrix elements (29) and (30) the interaction constant

$$\begin{aligned} G^{(1)}(k,k_1) &= \frac{\hbar|G_1(q)|^2(\omega_{\mathbf{q}} - \chi_{\mathbf{q}})}{[\hbar(\omega_{\mathbf{q}} - \chi_{\mathbf{q}})]^2 - [\varepsilon(\mathbf{k}_1 + \mathbf{q}) - \varepsilon(\mathbf{k}_1)]^2} \\ &\times \left[1 - \frac{2|G_2^b(\mathbf{q}, -\mathbf{q})|n\cos[2\varphi_1 - \varphi_b]}{\hbar(\omega_{\mathbf{q}} - \chi_{\mathbf{q}})}\right] \\ &\times \{[1 + \sum_{q'} f_{q'}^2 N_{q'}/\alpha]\}. \end{aligned}$$

for the elastic scattering process $\varepsilon_1(k_1) = \varepsilon_1(k)$ and $q \sim 2k_F$, $\alpha = \omega_q/|\kappa_q|$, $\kappa_q = |\tilde{A}|\omega_D f_q^2$ and $f_{q'}^2 = aq'/2$, can be approximately estimated by

$$\begin{aligned} G^{(1)} &\approx \frac{E_{11}}{2\rho c^2}\left(1 + \frac{2n|M_{12}|^2\cos\varphi_s}{2\hbar ck_F\varepsilon_{21} - 2n|M_{12}|^2\cos\varphi_s}\right) \\ &\times \left(1 - \frac{2|M_{12}|^2 n\cos[2\varphi_1 - \varphi_b]}{2\hbar\varepsilon_{21}ck_F - 2n|M_{12}|^2\cos[2\varphi_1 - \varphi_b]}\right) \\ &\times [1 + F(T)], \end{aligned}$$

where $M_{12} = \varepsilon_{21}p_{21}/\sqrt{\rho ck_F\hbar}$, $F(T) = |\tilde{A}|\omega_D a^2\sum_{q'}\omega_{q'}N_{q'}(T)/(2c)^2$ is anharmonic contribution.

Following the estimation of the expression for the mean energy of oscilation modes in the solid [44]

$$\hbar\sum_{q'}\omega_{q'}N_{q'}(T) = \langle E\rangle = 3Nk_BT\, D\left(\frac{\hbar\omega_D}{k_BT}\right), \tag{69}$$

where $D(x)$ is the Debye function. In accordance with the case: $\varphi_s \approx \varphi_b = 0$; and the argument of $G_1(q)$ is $\varphi_1 = \arg i = \pi/2$. In this case $G^{(1)}$ takes the maximal value . From this expression follows that $F(T) = 3Nk_BT|\tilde{A}|\omega_D a^2 D\left(\frac{\hbar\omega_D}{k_BT}\right)/(\sqrt{\hbar}2c)^2$. Taking into consideration that the number of atoms N enters in the expressions (6) and the renormalized anharmonic constant is invers proportional to this number $\tilde{A} = 3\hbar A/(KN\sqrt{KM})$, $\omega_D = 2\sqrt{K/M}$ we obtain the following expression $F(T) = 9k_BTAa^2/(2c^2KM) \times D\left(\frac{\hbar\omega_D}{k_BT}\right)$.

In the case when the anharmonic constant A takes the negative values in order to obtain the increasing of the critical temperature, we must take into consideration the next terms of perturbation theory. Indeed, considering that $A<0$ we can obtain the following expression for the re-normalization function, $\tilde{F}(T)$

$$\tilde{F}(T) = - |\tilde{A}|\omega_D a^2 \sum_{q'} \omega_{q'} N_{q'}(T)/(2c)^2 + a^3\omega_q^2 \tilde{B}\omega_D/(8c^3) \sum_{q'} \omega_{q'} N_{q'} \sum_{q''} \omega_{q''} N_{q''},$$

where the second term is obtained from interaction part of the Hamiltonian (24). Taking into consideration that the constant $\tilde{B}$ must be inverse proportional to the number of elementary cells of the crystal $\tilde{B} = B/N^2$, we can use the expression (69) for the simplification of new re-normalization function

$$\tilde{F}(T) = -9k_B T A a^2/(2c^2 KM) \times D\left(\frac{\hbar\omega_D}{k_B T}\right) + 2a^3 B\sqrt{K/M}\left[3k_B T\, D\left(\frac{\hbar\omega_D}{k_B T}\right)\right]^2 /(8c^3).$$

For the argument $2\varphi_1 - \varphi_b = \pi$ the constant $G^{(1)}$ Takes the minimal value. In the case, when $n|M_{12}|^2/\varepsilon_{21} >> 2\hbar c k_F$, the interaction constant takes minimal value, $G^{(1)} \to 0$. In this case it is necessary to take into account the Hamiltonian (63). After a simple modification of the effective Hamiltonian (63), in the first term we replaced $q+q_1 \to q$, $k_1 \to -k_1$ and used $q = k_1 - k$, whereas in the second term we replaced $q - q_1 \to q$, the following effective interaction Hamiltonian between the electrons is obtained

$$\check{H}_{Ieff2} = -\frac{G(T)}{V} \sum_{k,k_1} G_2(k,k_1) \check{a}^\dagger_{k1} \check{a}^\dagger_{-k_1} \check{a}_{-k} \check{a}_k, \tag{70}$$

where $k = (\mathbf{k}, \frac{1}{2})$, $-k = (-\mathbf{k}, -\frac{1}{2})$. For elastic scattering process $\varepsilon_1(k_1) = \varepsilon_1(-k)$ and $q \sim 2k_F$ it is obtained the following expression for interaction constant

$$\begin{aligned} G_2(k,k_1) &\approx G_2(T) = \frac{4}{V}\sum_{\mathbf{q}_1} \frac{|g_{21}(\mathbf{q}-\mathbf{q}_1)|^2 |g_{21}(\mathbf{q}_1)|^2}{\varepsilon_{21}^2} \\ &\times \left[\frac{1}{\hbar\omega_{\mathbf{q}-\mathbf{q}_1} + \hbar\omega_{\mathbf{q}_1}} + 4N_{\mathbf{q}_1} \frac{\hbar\omega_{\mathbf{q}-\mathbf{q}_1}}{(\hbar\omega_{\mathbf{q}-\mathbf{q}_1})^2 - (\hbar\omega_{\mathbf{q}_1})^2}\right]. \end{aligned} \tag{71}$$

According to the collective interaction Hamiltonian paths, (68) and (70), we obtain the following contribution to the potential of the Cooper pairing electrons in Hamiltonian of the superconducting system

$$\begin{aligned} \check{H} &= \sum_k (\varepsilon_1(k) - \mu)\check{a}^\dagger_k \check{a}_k - \frac{G(T)}{V}\sum_{k,k_1} \check{a}^\dagger_{k_1}\check{a}^\dagger_{-k_1}\check{a}_{-k}\check{a}_k, \\ G(T) &= G^{(1)} + G_2(T). \end{aligned} \tag{72}$$

Introducing the order parameter $\Delta(T) = \frac{G(T)}{2V}\sum_{k_1}\left\langle \check{a}^\dagger_{k_1}\check{a}^\dagger_{-k_1}\right\rangle$ and $\varepsilon_1'(\mathbf{k}) = (\varepsilon_1(\mathbf{k}) - \mu)$, we can approximate the Hamiltonian (72) in the following form:

$$\check{H} = \sum_k \varepsilon_1'(\mathbf{k})\check{a}^\dagger_k \check{a}_k - \sum_k (\Delta(T)\check{a}_{-k}\check{a}_k + h.c.). \tag{73}$$

Let us examine below the contribution of each mechanism of nonlinearity to the order parameter $\Delta(T)$. Considering that g_{21} takes zero value, when the distance between the first and second bands increase, we can take into a consideration only anharmonic contribution. In another case when the anharmonic coefficient takes zero value, we can study only nonlinearity connected to the two-photon transitions through the virtual band as this was studied in the papers [19], [43]. Below we divide these two mechanisms

I. Anharmonic nonlinear mechanism. The similar ideas of renormalisation of the phonon dispersion law by anharmonic correlates with the ideas of other Refs. [11, 37]. The authors of Ref.[11] proposed to use three forms of interaction of electrons with the vibrations of CuO_2 layers in the material $YBa_2Cu_3O_7$. One of this type takes into consideration the perturbation the nearest-neighbour pairs of $O4$ and $O2$, $O3$ ions; the second Hamiltonian part takes into consideration the energy changes in the $O2$ (O3) site due to the displacement of the plane oxygen ions in the crystal field formed due the charge difference of the Y^{3+} and Ba^{2+} ions. The third type of interaction model induced by the dimpling of the CuO_2 layers, which allows a modulation of the hopping matrix elements $t(R,R|r,r)$ by the Z-axis displacements of $O2$ and $O3$ ions. The last modifications of the matrix elements are similar with the renormalisation of the phonon operators in the CuO_2 layers under the influences of the vibrations in the Z direction. The problem which appears here consists in the representation of the order parameter and anisotropy of electron-phonon through the interaction constant. In this chapter, we avoid to present a quantitative model of this anisotropy considering that the dispersion law of the vibration in one of the directions is renormalised by the other two degrees of freedom of vibration modes.

Passing from the sun to integrals in the expressions for the order parameter as this is described in the literature [45, 17]

$$\ln\frac{\Delta(0)}{\Delta(T)} = 2\int_0^\infty \frac{dx}{\sqrt{x^2+(\Delta(T)/T)^2}[\exp\sqrt{x^2+(\Delta(T)/T)^2}+1]} - \frac{F(T)}{\rho[1+F(T)]}, \qquad (74)$$

where $\rho = G^{(1)}mk_F/(2\pi^2\hbar^2)$. We don't know exactly the value of the anharmonic parameter A, but we can represent the nonlinear expression through Debye function $F(T) = \Omega k_B T/(\hbar\omega_D) \times D(\hbar\omega_D/k_B T)$ in which the coefficient is included in $\Omega = 9Aa^2\hbar/(c^2M\sqrt{KM})$. In the Figure 7 is plotted the relative order parameter as a function of temperature and this relative coefficient Ω. As follows from the Figure 7 with increasing the coefficient Ω the order parameter increases too. Achieving the maximal value, it decreases till critical temperature in comparison with BCS theory. Let us take into consideration the negative value of anharmonic parameter $\tilde{A}$. In this situation instead of the function $F(T)$ in the expression (74) we must substitute it with $-F(T)$. According with the representation of the operators of the $su(2)$ symmetry (9) these negative an-harmonic behavior of the system diminishes the value of Debye frequency and the critical temperature too. The plot of Figure 8 demonstrates, that with increasing of the negative value of anharmonic parameter $\tilde{A}$ the critical temperature decreases too.

Following the idea of high an-harmonic behavior of High temperature superconducting materials, we propose to take into consideration the next expansion described by interaction

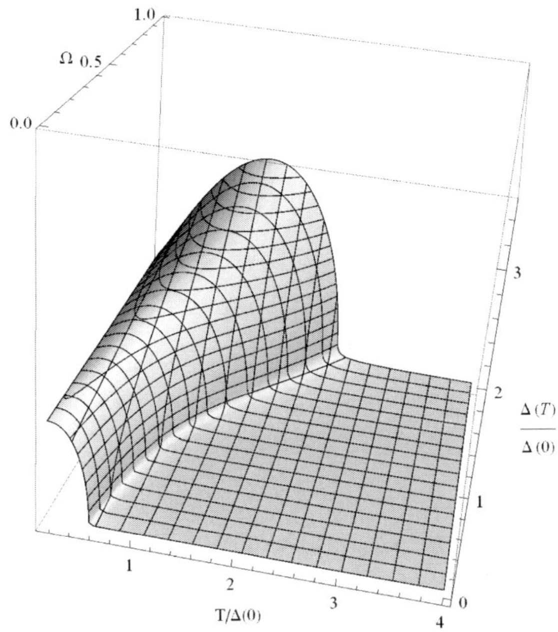

Figure 7. The relative order parameter for superconductivity $\Delta(T)/\Delta(0)$ as a function of relative temperature $(k_B T)/\Delta(0)$ and an-harmonic parameter Ω for following parameter values $\rho = 0.47, \hbar\omega_D/\Delta(0) = 0.6$.

part of the Hamiltonian (24) we obtain the following expression for non-linear modification of the order parameter

$$\tilde{F}(T) = -\Omega\left(\frac{k_B T}{\hbar\omega_D}\right) D\left(\frac{\hbar\omega_D}{k_B T}\right)\left[1 - E\left(\frac{k_B T}{\hbar\omega_D}\right) D\left(\frac{\hbar\omega_D}{k_B T}\right)\right],$$

where the new parameters are introduced $E = B\hbar a K^2/(c\hbar|A|)$ and $\Omega = 9|A|a^2\hbar/(c^2 M\sqrt{KM})$. In Figure 9 The time dependence on the parameter E is plotted. As follows from this plot the high non-linearity have the tendency to re-renormalise the elastic interaction between the atoms. In this situation the re-normalization function $\tilde{F}(T)$ of Debye frequency $\tilde{\omega}_D = \omega_D(1 - \tilde{F}(T))$, may pass from the negative value to positive one with increasing the temperature. For large values of parameter E the nuclei are constrained to occupy another equilibrium position and to restart the vibrations and the electron phonon interaction in the new minimum of collective potential function.

II. Non-linear two-quantum transitions between the electronic states. Considering that anharmonic coefficient takes zero value, we can find another non-linear mechanism of formation of Cooper pairs which involves the two-quantum cooperative process of interaction between the electrons belonging to a same conduction band. Indeed, representing the sum through the integral over q_1 and considering that dispersion law is linear $\omega_q = cq$ (the acoustic phonons), we obtain the following expression for the first term in the expression

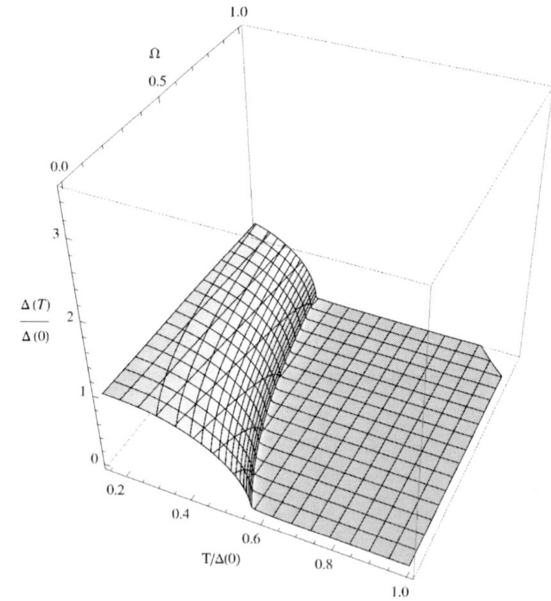

Figure 8. The decreasing of the relative order parameter for superconductivity $\Delta(T)/\Delta(0)$ with increasing of absolute value of negative an-harmonic parameter $A<0$ as a function of relative temperature $(k_BT)/\Delta(0)$ and coefficient Ω for the same values of the parameters as in Figure 7.

(71) term in (71)

$$\begin{aligned} I_2 &= \frac{4}{V}\sum_{\mathbf{q}_1}\frac{|g_{21}(\mathbf{q}-\mathbf{q}_1)|^2\,|g_{21}(\mathbf{q}_1)|^2}{\varepsilon_{21}^2}\frac{1}{\hbar\omega_{\mathbf{q}-\mathbf{q}_1}+\hbar\omega_{\mathbf{q}}}r \\ &= \frac{4V}{(2\pi)^3}\frac{1}{V\varepsilon_{21}^2}\frac{1}{[2\rho\hbar c]^2}[\varepsilon_{21}p_{21}]^2\int_0^{q_D}q_1^2dq_1\int_0^{2\pi}d\phi\int_0^{\pi}.\sin\theta d\theta \\ &\times \frac{1}{\hbar c}\frac{1}{|2\mathbf{k}_F-\mathbf{q}_1|+q_1}\times\frac{1}{q_1|2\mathbf{k}_F-\mathbf{q}_1|} \end{aligned} \tag{75}$$

Choosing the angle θ between the direction of the vectors $\mathbf{q}_1$ and $2\mathbf{k}_F$ and considering that that $2k_F << q_D$ we can represent the integral (75) in the following approximate form

$$\begin{aligned} I_2 &= \frac{p_{21}^2}{4\pi^2\rho^2\hbar^3c^3}[2\int_{2\kappa_F}^{q_D}dq_1\frac{1}{q_1}-\int_0^{\pi}.\sin\theta d\theta\int_{2\kappa_F}^{q_D}dq_1\frac{1}{|2\kappa_F+\mathbf{q}_1|+q_1}] \\ &\approx \frac{p_{21}^2}{4\pi^2\rho^2\hbar^3c^3}\ln[\frac{q_D}{2\kappa_F}]. \end{aligned}$$

In the similar way we obtain the following expression for interaction constant second

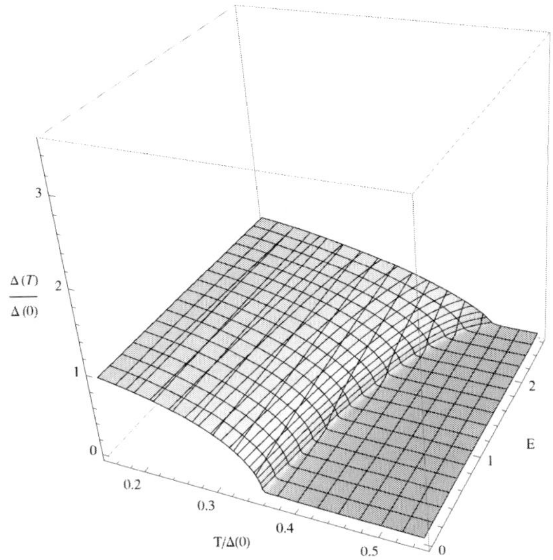

Figure 9. Possible restoration of the order parameter $\Delta(T)/\Delta(0)$ by hair order vibrations plotted as function of third order nonlinear parameter E and relative temperature $(k_BT)/\Delta(0)$. We Figure 7.

term in the expression (71)

$$\begin{aligned} IN_2 &= \frac{16}{V}\sum_{\mathbf{q}_1}\frac{|g_{21}(\mathbf{2k}_F-\mathbf{q}_1)|^2\,|g_{21}(\mathbf{q}_1)|^2}{\varepsilon_{21}^2}N_{\mathbf{q}_1}\frac{1}{\hbar c}\frac{|\mathbf{q}-\mathbf{q}_1|}{(\mathbf{2k}_F)^2-4\mathbf{k}_F q_1\cos\theta} \\ &= \frac{p_{21}^2}{4\pi^2\rho^2\hbar^3c^3k_F}\int_{k_F}^{q_D}N_{q_1}dq_1\int_0^{\pi}q_1\sin\theta d\theta\frac{1}{\kappa_F-q_1\cos\theta}. \end{aligned}$$

Using the Taylor series for the phonon number

$$N_{q_1}=\sum_{n=1}^{\infty}\exp\left(-n\frac{\hbar c q_1}{\kappa T}\right)$$

this integral takes the value

$$\begin{aligned} &\int_0^{q_D}dq_1\exp\left(-n\frac{\hbar c q_1}{\kappa_b T}\right)\{\ln|\kappa_F+q_1|-\ln|\kappa_F-q_1|\}= \\ &-\ \frac{\kappa_b T}{\hbar c n}\exp\left(n\frac{\hbar c\kappa_F}{\kappa_b T}\right)\left[Ei[-n\frac{\hbar c\kappa_F}{\kappa_b T}]-Ei[-n\frac{\hbar c(q_D+\kappa_F)}{\kappa_b T}]\right] \\ &+\frac{\kappa_b T}{\hbar c n}\exp\left(-n\frac{\hbar c\kappa_F}{\kappa_b T}\right)\left[Ei[n\frac{\hbar c\kappa_F}{\kappa_b T}]-Ei[-n\frac{\hbar c(q_D-\kappa_F)}{\kappa_b T}]\right] \\ &-\frac{\kappa_b T}{\hbar c n}\left[\exp\left(-n\frac{\hbar c q_D}{\kappa_b T}\right)\left[\ln(q_D+\kappa_F)-\ln(q_D-\kappa_F)\right]\right.. \end{aligned}$$

Here $Ei(x) = \int_{\infty}^{x} dy \exp[y]/y$ is the exponential integral [46]. For $q_D > k_F$ the temperature dependent term (71) takes the form

$$IN_2 \approx \frac{p_{21}^2}{4\pi^2\rho^2\hbar^3c^3k_F}\sum_{n=1}^{\infty}\frac{1}{n\zeta}\{\exp(-n\zeta)Ei(n\zeta) - \exp(n\zeta)Ei(-n\zeta)\},$$

an interaction constant can be represented through first and second order terms, where $\zeta = \hbar c k_F/(\kappa_B T)$.

$$G(T) = G^{(1)} + I_2(1+\Gamma),.$$

Here

$$\begin{aligned}\Gamma &= \left(\zeta \ln\left|\frac{q_D}{k_F}\right|\right)^{-1}\sum_{n=1}^{\infty}\frac{1}{n}\{\exp(-n\zeta)Ei(n\zeta) - \exp(n\zeta)Ei(-n\zeta)\},\\ \zeta &= \hbar c k_F/(\kappa_B T).\end{aligned}$$

In the limit $\zeta \gg 1$ the integral exponential function can be approximated by $\Re Ei(x) \sim \exp(x)/x$. In this case it is obtained the following expression for parameter $\Gamma = \varphi T^2$, where $\varphi = 2[\kappa_B/(\hbar c k_F)]^2(\ln|q_D/k_F|)^{-1}$.

After diagonalization of Hamiltonian (73) the Bogolyubov transformations [47] the following expression for the order parameter as a function of temperature is obtained:

$$\Delta(T) = \frac{G(T)}{2V}\sum_k \frac{\Delta(T)}{\sqrt{\eta_k^2 + \Delta^2(T)}}\tanh\frac{\sqrt{\eta_k^2+\Delta^2(T)}}{2k_bT}, \tag{76}$$

where $\eta_k = \varepsilon_1(k) - \mu$. Moreover we can pass from the sum in k - space to an integral over η_k (see [45, 48, 49]). In the Debye layer the sum is approximated by integral

$$\sum_k \ldots = \frac{V}{2\pi^2}\int k^2 dk \ldots \approx V\frac{m^* k_F}{2\pi^2\hbar^2}\int_0^{\omega_D} d\eta_k \ldots,$$

so that we obtain the following equations for order parameter

$$\begin{aligned} &for\ \ T \neq 0 && \frac{1}{G(T)N(0)} = \int_0^{\omega_D}\frac{d\eta}{\sqrt{\eta^2+\Delta^2(T)}}\\ & && - \ 2\int_0^{\infty}\left(\exp\frac{\sqrt{\eta^2+\Delta^2(T)}}{k_bT}+1\right)\frac{d\eta}{\sqrt{\eta^2+\Delta^2(T)}},\\ &for\ \ T = 0 && \frac{1}{G(0)N(0)} = \int_0^{\omega_D}\frac{d\eta}{\sqrt{\eta^2+\Delta^2(0)}},\end{aligned} \tag{77}$$

where $G(0) = G^{(1)} + I_2$, $N_F(0) = m^* k_F/(2\pi^2\hbar^2)$ is the density of states on the Fermi surface. Taking into account that the first integral in (77) in these expressions can be represented by expression $arcsih(\omega_D/\Delta) \approx \ln[2\omega_D/\Delta]$, we obtain the following expression for the differences of two equations

$$\ln\frac{\Delta(0)}{\Delta(T)} = 2I\left(\frac{\Delta(T)}{k_bT}\right) - \frac{\Gamma}{\lambda(1+\delta+\Gamma)}, \tag{78}$$

where

$$I(u) = \int_0^\infty \frac{dx}{\sqrt{x^2+u^2}(\exp\sqrt{x^2+u^2}+1)}.$$

Here, we assume that $\Delta(T)$ is constant in the Debye layer $2\hbar\omega_D$ around the Fermi surface. $\lambda = \left|G^{(1)}(T)+I_2\right|N_F(0)$; $\delta = I_2/G^{(1)}(T)$; $\Delta(0) = 2\hbar\omega_D\exp(-1/\lambda)$. Using the integral expression of the MacDonald's function, we can represent the equation (78) in the following form

$$\ln\frac{\Delta(0)}{\Delta(T)} = 2\sum_{n=1}^{\infty}(-1)^{n+1}K_0\left(n\frac{\Delta(T)}{k_bT}\right) - \frac{\Gamma}{\lambda(1+\delta+\Gamma)}, \quad (79)$$

where $K_0(u)$ is the MacDonald's function. We observe that for low values of temperature $(T/\Delta(0) << 1)$ our proposed model differs substantially from the BCS-model with constant exchange integrals between the carriers. In Figure 3 the difference between the investigated model and BCS-model are plotted for two cases. We can divide this process in two stages A and B.

A. The increasing order parameter at low temperature can be analytically described if we solve the equation (79) in the limits $\Delta(T) > k_bT$. In this case it is assumed that $K_0(u) \sim \sqrt{\pi/2u}\exp(-u)$ for $u \to \infty$, so it is sufficient to take the term of the series corresponding to $n = 1$. As the order parameter $\Delta(T)$ is close to $\Delta(0)$, we can approximate $\ln\Delta(0)/\Delta(T) \approx [\Delta(0) - \Delta(T)]/\Delta(0)$. In the region of low temperature the following temperature dependence of the order parameter is obtained:

$$\Delta(T) = \Delta(0) - \sqrt{2\pi k_bT\Delta(0)}\exp\left(-\frac{\Delta(0)}{k_bT}\right) + \frac{\Delta(0)}{\lambda}\frac{\varphi T^2}{1+\varphi T^2}. \quad (80)$$

For a small temperature, it follows from expression (80) that the order parameter, $\Delta(T)$, increases with increasing temperature (see Figure 10). This behavior of the order parameter differs from BCS single-phonon theory. At low temperatures one observes an increase of the order parameter. After achieving the maximum value, the order parameter decreases (see Figure 10). This phenomenon is connected with the strong dependence of the exchange integral on temperature.

B. In the critical point, $T = T_c$, the order parameter takes the value zero. Considering the condition $\Delta(T_c) = 0$ in equations (76) and (78) one can obtain the following equation for the critical temperature:

$$\ln\left(\frac{\pi k_bT_c}{\Delta(0)}\right) = \frac{1}{\lambda}\frac{\varphi T_c^2}{1+\varphi T_c^2} + C_0. \quad (81)$$

Here $C_0 = 0.577$ is the Euler constant. From equation (81) it follows that the intersection of the logarithm $\ln(aT)$ with the function $\varphi T^2/(\lambda+\lambda\varphi T^2)$ gives us the critical temperature (here $a = \exp(-C_0)\pi k_b/\Delta(0)$). The main difference between the BCS-like and our proposed model consists in the influence of parameters φ and $1/\lambda$ on the critical temperature. For numerical comparison, we use the BCS-like model with $\varphi = 0$ in equation (81). In this case the well known connection between critical temperature and the gap is obtained. In the model of this paper the critical temperature increases when the parameter φ increases. But this influence of φ on the critical temperature in comparison with the BCS-like model can be observed for small values of the parameter λ, which is equivalent

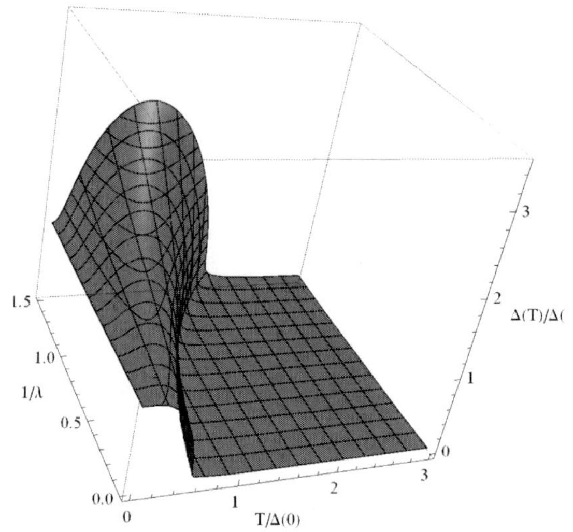

Figure 10. The relative order parameter for superconductivity versus temperature and non-linear $1/\lambda$. The other interaction parameters: $\hbar ck_F/\Delta(0) = 0.3$, $\delta = 0$ as in Figure 3.

to small values of $\Delta(0)$. In the Figure the difference between the BCS-like and proposed models (with $\varphi \neq 0$) is represented. As we observe, the behavior of our model in comparison with BCS-like being more evident for a small gap. In order to estimate the power law dependence of the order parameter near the critical temperature (T_c) one can expand the integral in equation (78) on the small parameter $\Delta(T)/(k_b T)$. Taking into account the equations (76)-(79) for T_c, we obtain the following temperature dependence of the order parameter

$$\Delta(T) = \frac{4\pi k_b T_c}{\sqrt{14\xi(3)}} \left[\left(1 - \frac{T}{T_c}\right) \left(1 - \frac{2\varphi T_c^2}{\lambda(1+\varphi T_c^2)}\right) \right]^{1/2}. \tag{82}$$

where $\xi(3)$ is the value of Riemann Zeta-function ($\xi(3) \approx 1.2$).

he decrease of $\Delta(T)$ as the function of temperature is slower in the proposed model than in the BCS-model. This follows from numerical results and the analytical dependence (82). Indeed the absolute values of the temperature derivatives of the order parameter are less than in the BCS-model, $|\ \partial\Delta(T)/\partial T\ | < |\ \partial\Delta^{BCS}(T)/\partial T\ |$. The slow decrease of the parameter $\Delta(T)$ gives us a higher critical temperature than in traditional superconductivity theory with constant exchange integral when $\sigma = 0$.

Two models of the re-normalization of interaction between potential the electrons are proposed for the representation of temperature dependence of the order parameter in the superconductivity phase. These combinations of these models open the new possibilities in the improving of the concept of non-linear superconductivity. For example the selective excitation of the superconductor with the acoustic wave may be observed studying the dependence of the critical temperature as a function of the intensity of this wave and its quantum statistical properties.

6. Discussions and Possibilities of Application of Non-linear Superconductivity

In this chapter, we discussed the effects connected with an-harmonic lattice vibration and nonlinear transitions. The combination between the nonlinear effect, stimulated by renormalisation of the frequencies of the vibration modes, and multi-phonon transition effect was taken in consideration in our proposed approach. These nonlinear effects were studied in the Ref.[35] for intensity coupling vibration modes in solids and multi-phonon aspects of interactions between the electrons proposed earlier in the papers [19] and [43]. This chapter has taken into account both non-linearity in the effective Hamiltonian (62), (63) and (67). We are interested in the contribution of non-linearity connected to the description of non-linear lattice vibrations and the re-normalization of vibration frequencies. We observed that the interaction Hamiltonian (62), which corresponds to the positive an-harmonic parameter $A > 0$ this approach gives the increasing of the Debye frequency and coherence between the Cooper pairs. This effect, brings the increasing of the critical temperature of the superconductivity state (see Figures 7, 10). The nonlinear lattice vibration corresponds to $su(1,1)$ algebra and with the excitation of oscillation mode k, the frequency of nonlinear oscillator increase too $\omega_q[1+\sum_{q'} f_{q'}^2 N_{q'}(1+\delta_{q,q'})/\alpha]$. This temperature re-normalization of the elastic interaction constant between the nuclei becomes possible, studying the vibrations of B and CuO_2 layers in high temperature superconducting materials like MgB_2 and $YBa_2Cu_3O_{6.5}$.

In opposite case when $\kappa_k < 0$ with increasing of the temperature the interaction constant in the Hamiltonian (62) decrease. This effect is accompanied with decreasing of the frequency of the nonlinear oscillators $\omega_q[1-\sum_{q'} f_{q'}^2 N_{q'}(1+\delta_{q,q'})/\alpha]$ and we plotted the similar scenario in relative units of the system parameters in Figure 8. In this situation the increasing of critical temperature may be achieved, studying higher order non-linear terms in the of the lattice vibration Figure 9.

Another result of this chapter discussed the competition between one- and the two-quantum exchanges between electrons or atoms through the boson field. In particular, it is shown that the interference between the first and the second order electron-boson interactions can be observed. This attractive situation can be achieved, when the second order exchange between the carriers drastically diminishes the zero temperature value of the order parameter. It was shown that the exchange integral between the electrons increases with temperature, due to the fact that the two-quantum scattering effect stimulates the coherent formation of Cooper pairs in superconductivity. This effect opens a new concept of the formation of Cooper pairs in the processes of multi-photon exchanges between the current carriers.

The proposed models give us the possibility to take into account the cooperative interaction between the cooper pairs through the non-linear lattice vibration in single and two-phonon exchange processes between the electrons. The similar problem, it is proposed for quasi-spin systems used in the description of super-radiance and ferro-magnetism [43]. These effects of the temperature modification of critical points in superconducting materials open the new possibilities of the manipulation with superconducting gap. The external excitation of the vibration modes in such materials follows from the nature of its nonlinear proprieties analyzed in the literature. The peculiarities of the non-linear vibrations may be significant in the explanation of several experimental realizations [7, 8, 50], in which the quadratic electron-phonon coupling gives rise to the frequency dependence of the order parameter.

References

[1] Bardeen, J; Cooper, J. & Schrieffer, J. (1957).Theory of Superconductivity. *Phys. Rev.* 108(5), 11751204.

[2] Ngai, K. L. (1974). Two-Phonon Deformation Potential and Superconductivity in Degenerate Semiconductors. *Phys. Rev. Lett.* 32, 215218.

[3] Bloch, C. & De Dominicis, C. (1958). Un dveloppement du potentiel de gibbs dun systeme quantique composd un grand nombre de particules. *Nucl. Phys.* 7, 459-479.

[4] Bednorz, J. G. & Muller, K. A. (1986). Possible high Tc superconductivity in the BaLaCuO system. *Z. Phys. B* 64, 189-193.

[5] Brizhik, L. S. & Davydov, A. S. (1984). The Bisoliton Model of High-T c Superconductivity. *Fiz. Nizk. Temperatur*, 10 358-365.

[6] Enaki, N.A., Prepelitza O.B. (1992). The specific of the creation of Couper pairs in two-phonon exchanges with thermostat. *Izvestia Academii Nauk*, Rep.of Moldova 3(9), 52-58.

[7] Farst, M., Manzoni,C. ; Kaiser, S.; Tomioka, Y.; Tokura, Y.; Merlin, R. & Cavalleri A. (2011). Nonlinear phononics as a new ultrafast route to lattice control. *Nature Physics*, 7, 854-856.

[8] Farst, M.; Mankowsky, R.; Bromberger, H. , Fritz, D.M.; Lemke, H.; Zhu, D.; Chollet, M., Tomioka, Y., Tokura, Y., Merlin, R.; Hill, J.P.; Johnson, S.L. & Cavalleri, A. (2013). Displacive lattice excitation through nonlinear phononics viewed by femtosecond X-ray diffraction, *Solid State Communications*, 169 24-27.

[9] Gluhoi A. Yu., Neghinskii I.V. (2013). Equipment for automatic investigation of nonlinear parameters High - Tc ceramic $YBa2Cu3O_{7-x}$ superconductors in the variable magnetic fields. *Journal Don Engineering Bulletin*, Russia, N 2 (25) / Vol 25 / 2013.

[10] Chubukov, A. V.; Pines, D.; Schmalian, J. *The Physics of Conventional and Unconventional Superconductors* edited by Bennemann, K. H.; Ketterson, J. B. (Springer-Verlag) preprint cond-mat/0201140.

[11] Sherman, E.Ya. & Ambrosch-Draxl C. (2002). Probing electron-phonon coupling in high. Tc superconductors by site-selective isotope substitution. *Eur. Phys. J.* B. 26, 323-328.

[12] Szabo, P.; Samuely, P. ; Kaccmarccik, J.; Klein, T.; Marcus, J.; Fruchart, D.; Miraglia, S.; Marcenat, C. & Jansen, A. G. M. (2001). Evidence for Two Superconducting Energy Gaps in MgB2 by Point-Contact Spectroscopy. *Phys. Rev. Lett.* 87(13), 137005-4p.

[13] Wang, Y; Plackowski, T & Junod, A. (2001) Specific heat in the superconducting and normal state (2-300 K, 0-16 T), and magnetic susceptibility of the 38 K superconductor MgB2: evidence for a multicomponent gap. *Physica C*, 355(3-4), 179-193.

[14] Bohnen, K.-P., Heid, R. & Renker, B. (2001). Phonon dispersion and electron-phonon coupling in MgB2 and AlB2. *Phys. Rev. Lett.* 86, 5771-5774.

[15] Yildirim, T. ; Gulseren, O.; Lynn, J. W.; Brown, C. M.; Udovic, T. J.; Huang, Q.; Rogado, N.; Regan, K. A.; Hayward, M. A.; Slusky, J. S.; He, T.; Haas, M. K.; Khalifah, P.; Inumaru, K. & Cava, R. J. (2001). Giant Anharmonicity and Nonlinear Electron-Phonon Coupling in MgB_2: A Combined First-Principles Calculation and Neutron Scattering Study. *Phys. Rev. Lett.* 87(3), 037001-4p.

[16] DiRienzo, A. & Rogovin, D. (1978). *Coherence in Spectroscopy and Modern Physics*, New York, Plenum Press.

[17] Abrikosov, A.A.; Gorkov, L.P. & Dzyaloshinskii, I.Ye. (1965). *Quantum Field Theoretical Methods in Statistical Physics*, Pergamon Press, Oxford ea, xii,365 pp.

[18] Levich, V. G. & Vdovin, Yu. A. (1962) *The Course of Theoretical Physics*, Moscow, Nauka.

[19] Enaki, N.A. & Eremeev, V. (2002). Two-phonon cooperative phenomena in superconductivity, *New Journal of Physics* 4, 80.1-80.12.

[20] Enaki, N. A. (1988). Superradiation from two-photon spontaneous decay. *Sov. Phys. JETP*, 67, 2033-2038; [(1988) *Zh. Eksp. Teor. Fiz.*, 94, 135].

[21] Bak Z. (2004). Coupling of Electrons to the Multiphonon Complexes and its Relevance to Superconductivity. *Czechoslovak Journal of Physics*, 54(4), D461-D464, Part 2, Suppl.

[22] Dubrovski, O. A. & Orlov, A. V. (1994). Coupled biphonon states of acustic phonons. *Fiz. Tverd. Tela* (St. Petersburg) 36, 3131; (1994). *Phys. Solid State*, 36, 1663.

[23] Page, J. B. (1990), Asymptotic solutions for localized vibrational modes in strongly anharmonic periodic systems. *Phys. Rev. B* 41(11), 78357838.

[24] Pang, X. F., Feng Y. P. (2005), *Quantum Mechanics in Nonlinear Systems*, NY, World Scientific.

[25] Andersen, O.K.; Liechtenstein, A.I.; Rodriguez, O.; Mazin, I.I.; Jepsen, O.; Antropov, V.P.; Gunnarsson, O. & Gopalan, O. (1991). Electrons, phonons, and their interaction in YBa2Cu3O7. *Physica C*, 185-189, 147-155.

[26] Andersen, O.K.; Liechtenstein, A.I.; Jepsen, O.; Paulsen F. (1995). LDA energy bands, low-energy hamiltonians, t?, t?, $t_{\perp}(k)$, and $J_{\perp}$. *J. Phys. Chem. Solids* 56, 1573-1591.

[27] Jepsen, O.; Andersen, O.K.; Dasgupta, I. & Savrasov, S.Y. (1998). BUCKLING AND d-WAVE PAIRING IN HiTc SUPERCONDUCTORS. *J. Phys. Chem. Solids* 59, 1718-1722.

[28] Makhankov, V. G. & Fedyanin, V. K. Non-linear effects in quasi-one-dimensional models of condensed matter theory. (1984) *Phys. Rep.* 104(1), 1-86.

[29] Anselm, A. I. (1978) *Vvedenie v Teoriu Poluprovodnikov*, Moskow, Nauka.

[30] Holstein, T. & Primakoff, H. (1940). Field Dependence of the Intrinsic Domain Magnetization of a Ferromagnet. *Phys. Rev.* 58(12), 1098-1113.

[31] Bargmann V., (1961). On a Hilbert space of analytic functions and an associated integral transform part I. *Communication on Pure and Appl. Math.* 14(3), 187-214.

[32] Davidov, A. (1976). *Teoriya Tverdovo Tela*, Moskaw, Nauka.

[33] Bjorken, J.D. & Drell, S.D. (1964). *Relativistic Quantum Mechanics*, (MGH).

[34] Itzykson, C. & Zuber, J.-B. (2006). *Quantum Field Theory*, Dover Books on Physics.

[35] Aynajian P. (2010). *Electron-Phonon Interaction in Conventional and Unconventional Superconductors*. Berlin-Heidelberg, Springer-Verlag.

[36] Buzea Cristina & Yamashita Tsutomu (2001). *Review of superconducting properties of* MgB_2, cond-mat/0108265, 35 p.

[37] Crespi, V.H. & Cohen, M.L. (1993). Anharmonic phonons and high-temperature superconductivity. *Phys. Rev. B*, 48(1), 398406.

[38] Trotsenko S., Kumar A., Volotka A.V. et al. (2010) Spectral Shape of the Two-Photon Decay of the 2S0 State in He-Like Tin. *Phys. Rev. Lett.* 104(3) 033001-4p.

[39] Chluba J. and Sunyaev R. A. (2006). Time-dependent corrections to the Ly α escape probability during cosmological recombination. *Astronomy & Astrophysics* 446, 619-635.

[40] Blum, K. (1981) *Density Matrix Theory and Application*, New York, Plenum Press.

[41] Enaki, N.A. (1990). Photon statistics in two-quantum collective optical nutation. *Sov. Phys. JETP*, 71(3), 435-442 [Russian original - ZhETF, Vol. 98, No. 3, p. 783].

[42] Enaki N. A., Colun S. (2012). Nonlinear effects in the theory of superconductivity. *J. Phys.: Conf.Ser.* 338 012006.

[43] N.A. Enaki, Cooperative Two-Quanta Phase Transitions in Quantum Optics and Superconductivity. *Journal of Low Temperature Physics*. Vol. 160(5), 157-179 (2010).

[44] Landau L.D., Lifshitz E.M. 1980 *Statistical Physics*. 5 (3rd ed.). Butterworth- Heinemann. ISBN 978-0-7506-3372-7.

[45] Pitaevskii L. P., Lifshitz E.M. 2010. *Statistical Physics*, Part 2: Volume 9, Oxford, Elsevier Group.

[46] Janke E. Ja, Emde F., Losch F. 1960, *Tafeln Hoherer Functionen*, Stuttgart, 342 p.

[47] Bogoliubov N. N., Bogoliubov(j) N. N. 1980. *Fiz. Elem. Ciastits At. Yadra* 11, 245.

[48] Enaki,N.A & Eremeev, V. (2006). Specificity of Phase Transition of Quasi-Spin System in Two-Quantum Exchange with Thermostat. *Phys. Lett. A*, 357, 104-107.

[49] Enaki, N.A & Macovei, M.A. (1999). Two-photon cooperative decay in a cavity in the presence of a thermalized electromagnetic field. *Rus. Phys. JETP* [http://link.springer.com/article/10.1134/1.558839], 88, 633-641.

[50] Kaiser, S., Hunt, C. R., Nicoletti, D., Hu, W., Gierz, I. , Liu, H. Y., Le Tacon, M., Loew, T., Haug, D., Keimer, B. & Cavalleri A. (2014). Optically induced coherent transport far above Tc in underdoped YBa2Cu3O6+, *Phys. Rev. B* 89, 184516.

In: Superconductivity
Editor: Paulette Grant

ISBN: 978-1-63483-907-5

Chapter 4

GENERALIZED BOSE-EINSTEIN CONDENSATION MODEL OF SUPERCONDUCTIVITY AND SUPERFLUIDITY

***I. Chávez*[1]*,*[*] *J. Ortega*[1]*, M. de Llano*[1] *and V.V. Tolmachev*[2]**
[1]Instituto de Investigaciones en Materiales,
Universidad Nacional Autónoma de México, DF, México
[2]Baumann State Technical University, Moscow, Russia

Abstract

This survey begins with a detailed derivation of the three coupled integral equations associated with the so-called generalized Bose-Einstein condensation (GBEC) formalism of a many-fermion system. The formalism can describe superconductors if the fermions are either electrons or holes. Changing the two-fermion interactions allows description of neutral-fermion superfluids (such as liquid ^{3}He, neutron matter, or ultracold trapped Fermi gases like ^{40}K or ^{7}Li). In essence, the GBEC formalism accounts not for boson-boson interactions (as BCS theory also ignores) but for boson-fermion (BF) interaction vertices. It leads to two gap-like and a number equation. The three unknowns associated with the model are functions of temperature, coupling and initial fermion number-density, and are: the fermion chemical potential and the BEC number-densities of zero total (or center-of-mass) momentum of two-electron (2e) and two-hole (2h) Cooper pairs (CPs) that continually form and disintegrate within the many-fermion medium as a result of the vertex (two-fermion/one boson) interactions. A detailed sketch follows of how the three equations reduce in the appropriate limits to the five known continuum (as opposed to "spin" or "lattice") statistical models of superconductivity (and by extension to superfluidity), including BCS, BCS-Bose *crossover*, the Friedberg-T.D. Lee and BEC theories. It is also seen that by ignoring either 2h-CPs or 2e-CPs the GBEC cannot give the BCS temperature-dependent gap equation (nor, therefore, the BCS critical temperature). Although explicit mention of individual holes is unnecessary, both kinds of CPs are needed. Finally, practical consequences for 3D superconductors are reviewed which feature critical BEC temperatures T_c enhanced by several orders of magnitude over those of BCS theory, assuming the same interfermion interaction, namely the Cooper-BCS (four-fermion) interaction

[*]israelito@ciencias.unam.mx

model corresponding to the assumed BF vertex interaction. These enhancements in T_c result from the GBEC's ability to depart from the perfect 2e/2hCP symmetry to which BCS theory is restricted by construction.

PACS 74.20.-z; 74.20.Fg; 03.75.F

Keywords: generalized BEC, hole Cooper pairs, crossover BCS-Bose, high T_c superconductors

1. Introduction

Boson-fermion (BF) models of superconductivity (SC) as a Bose-Einstein condensation (BEC) [1, 2] go back to the mid-1950's [3–9] pre-dating even the BCS-Bogoliubov theory [10–17]. Although BCS theory only contemplates the presence of "Cooper correlations" [18] of single-particle states, BF models [7, 9], [19–34] posit the existence of actual bosonic CPs. Such pair charge carriers have been observed in magnetic flux quantization experiments on elemental [35, 36] as well as on cuprate [37] superconductors (SCs). Clusters larger than pairs, viz., *quartets* or quadruples with charge $4e$, have *not* been unambiguously observed in the bulk of any superconductor (see, however Refs. [38–42]). The presence of quartets in ^3He in aerogel [43] has been suggested. Moreover, no experiment has yet been done, to our knowledge [44], that distinguishes between electron and hole pairs, i.e., that determines the *sign* of $2e$ charge carriers. CPs appear to be the single most important universally accepted ingredient of SC, whether conventional or "exotic" and whether of low- or high-transition-temperatures T_c. And yet, in spite of their centrality they are poorly understood. The fundamental drawback of early [7, 9] BF models, which took two-electron (2e) pairs as analogous to diatomic molecules in a classical atom-molecule gas mixture, is the cumbersome introduction of an electron energy gap $\Delta(T)$. "Gapless" models cannot describe the ordinary superconducting state at all, although they are useful in locating transition temperatures if approached from above, i.e., $T > T_c$. Even so, we are not aware of any calculations with the early BF models attempting to reproduce any empirical T_c values.

The gap first began to appear naturally in later BF models [19–31]. With two [26, 27] exceptions, however, all BF models neglect the effect of *hole* CPs included on an equal footing with electron CPs to give the "complete" BF model that constitutes the generalized Bose-Einstein condensation (GBEC) theory to be surveyed. It is complete only in that it consists of *both* bosonic CP species coexisting with unpaired electrons.

For 13 or so years the highest T_c value for any superconductor was $23K$, until the discovery [45] in 1986 of the first so-called "high-T_c" superconductor (HTSC) $LaBaCuO$ having a $T_c \simeq 35K$. A feverish search for materials with even higher T_c's lead within just seven years to the highest-T_c cuprate superconductor known to date, the $HgBaCaCuO$ compound [46] with a $T_c \simeq 164\ K$ under very high pressure ($\simeq 310,000$ atm). Most recently, the discovery of the hydride SC compound H_3S [47] with $T_c \simeq 193K$ under ultrahigh pressure ($\simeq 2 \times 10^6$ atm) marks a new record in HTSC.

2. GBEC Theory

2.1. GBEC Hamiltonian

The generalized Bose-Einstein condensation theory (GBEC) describes in detail the properties of a quantum statistical many-fermion system. For this purpose we assume that our system has three kinds of individual micro-particles, unpaired particles (here, electrons), two-particle Cooper pairs (2e-CPs) and two-hole Cooper pairs (2h-CPs) both kinds of which are postulated to be bosons. In this system we have interactions between unpaired particles as well as between unpaired particles or holes with 2e-CPs or 2h-CPs. We omit the interactions between 2e-CPs and between 2h-CPs. We take into account very specific direct interactions between fermions and 2e-CPs and 2h-CPs due to elementary processes of decay of 2e- or 2h-CPs into their unpaired constituents of particles or holes, and the inverse elementary processes of formation of both kinds of pairs.

We first describe the *kinematic properties* of the many-fermion system. Assume that the many-fermion system is contained in a vessel of macroscopic dimensions of volume L^d where d is the system dimensionality. One-particle states of unpaired individual particles are described by plane-wavefunctions

$$\phi_{\mathbf{k},s}(\mathbf{r},\sigma_z) = L^{-d/2}e^{i\mathbf{k}\cdot\mathbf{r}}\,\delta_{s,\sigma_z} \tag{1}$$

which are characterized by values of the fermion momentum wavevector $\mathbf{k}$ and by its projection of spin $s = \pm\frac{1}{2}$ on z-axis. These states have energies

$$\epsilon_k \equiv \frac{\hbar^2 k^2}{2m} \tag{2}$$

where m is the effective fermion mass. One-particle states of the center-of-mass motion of individual 2e-CPs are described by the plane waves

$$\Phi_{\mathbf{K}}(\mathbf{R}) = L^{-d/2}\exp[i\mathbf{K}\cdot\mathbf{R}] \tag{3}$$

which are characterized by values $\mathbf{K}$ of the total, or center-of-mass momentum (CMM), wave-vectors of electron Cooper pairs (CPs). These states have energies

$$E_+(K) \equiv E_+(0) + \varepsilon_K \tag{4}$$

where $E_+(0)$ is the positive energy of *motionless* (i.e., zero-CMM) 2e-CP and ε_K is the non-negative excitation energy (or, dispersion relation) of the *moving* (i.e., with nonzero-CMM) CP. This would be simply $\varepsilon_K \equiv \hbar^2K^2/2(2m)$ if we assumed *incorrectly* that 2e-CPs with a mass of $2m$ moved *in vacuo* as does, say, a deuteron with an average nucleon mass of m. In fact, the original CPs propagating slowly in the Fermi sea are *linearly-dispersive*, namely,

$$\varepsilon_K = 2\Delta + b(d)\lambda\hbar v_F K + \mathcal{O}(K^2) \tag{5}$$

with $b(d) = 1, \frac{1}{2}\pi$ and $\frac{1}{4}$ for $d = 1$ [48], 2 [49] and 3 [50], respectively, Δ is the zero-temperature BCS energy gap, and λ a dimensionless interfermionic coupling constant, e.g., the Cooper/BCS model interaction coupling constant associated with pair formation. Note the notational difference between *fermion* (2) ϵ and *boson* (5) ε dispersion energies .

As the second quantization basic functions of one-particle states of center-of-mass motion of individual 2h-CPs, we also take the plane waves (3) but referring to CMM values $\mathbf{K}$ of 2h-CPs. These states have energies

$$E_{-}(K) \equiv E_{-}(0) - \varepsilon_K \tag{6}$$

where $E_{-}(0)$ is the positive energy of *motionless* 2h-CP. We assume here that the mass of the 2h-CP equals $2m$, exactly as for the 2e-CP. For simplicity, we also assume that a 2e-CP as well as a 2h-CP have only *one* discrete internal bound state.

Now let us use the following ordinary in the second quantization representation periodic boundary conditions on the walls of the vessel of volume L^d for all wave functions of our many-fermion system, which we consider. They are satisfied automatically when momentum $\mathbf{k}$ of individual fermions, and momentum $\mathbf{K}$ of individual 2e-CPs, and momentum $\mathbf{K}$ of individual 2h-CPs take those values which are the *nodes* of simple lattice in k-space (reciprocal space), the elementary cell of the lattice has volume $(2\pi/L)^d$ (Born-von Karman boundary conditions).

The second-quantized creation and annihilation operators of fermions in their one-particle states with momenta $\mathbf{k}$ and spin projection s are designated by $a^{\dagger}_{\mathbf{k},s}$ and $a_{\mathbf{k},s}$ These are Fermi operators, i.e., they satisfy the *anti*commutation relations

$$\begin{aligned}
\{a^{\dagger}_{\mathbf{k},s}, a^{\dagger}_{\mathbf{k}',s'}\} &\equiv a^{\dagger}_{\mathbf{k},s}a^{\dagger}_{\mathbf{k}',s'} + a^{\dagger}_{\mathbf{k}',s'}a^{\dagger}_{\mathbf{k},s} = 0, \\
\{a_{\mathbf{k},s}, a_{\mathbf{k}',s'}\} &= 0 \\
\{a_{\mathbf{k},s}, a^{\dagger}_{\mathbf{k}',s'}\} &= \delta_{\mathbf{k}',\mathbf{k}}\delta_{s',s}
\end{aligned} \tag{7}$$

where $\delta_{\mathbf{k}',\mathbf{k}}$ and $\delta_{s',s}$ are Kronecker symbols.

The second-quantized creation and annihilation operators of 2e-CPs in their one-*particle* states of CMM with wavevector $\mathbf{K}$ are designated by $b^{\dagger}_{\mathbf{K}}$ and $b_{\mathbf{K}}$. We *postulate* them to be Bose operators, i.e., that they satisfy the commutation relations

$$\begin{aligned}
[b^{\dagger}_{\mathbf{K}}, b_{\mathbf{K}'}] &\equiv b^{\dagger}_{\mathbf{K}}b_{\mathbf{K}'} - b_{\mathbf{K}'}b^{\dagger}_{\mathbf{K}} = 0 \\
[b_{\mathbf{K}}, b_{\mathbf{K}'}] &= 0 \\
\left[b_{\mathbf{K}}, b^{\dagger}_{\mathbf{K}'}\right] &= \delta_{\mathbf{K}',\mathbf{K}}.
\end{aligned} \tag{8}$$

Note that the last relation differs from the corresponding one associated with the *BCS pair operators* $b_{\mathbf{kK}} \equiv a_{\mathbf{k}_2,\downarrow}a_{\mathbf{k}_1,\uparrow}$ and $b^{\dagger}_{\mathbf{kK}} \equiv a^{\dagger}_{\mathbf{k}_1,\uparrow}a^{\dagger}_{\mathbf{k}_2,\downarrow}$ (for which only the $\mathbf{K} = 0$ case was discussed by them). That relation is

$$\left[b_{\mathbf{kK}}, b^{\dagger}_{\mathbf{k}'\mathbf{K}}\right] = (1 - n_{\mathbf{K}/2-\mathbf{k},\downarrow} - n_{\mathbf{K}/2+\mathbf{k},\uparrow})\delta_{\mathbf{kk}'} \tag{9}$$

where $n_{\mathbf{K}/2\pm\mathbf{k},s} \equiv a^{\dagger}_{\mathbf{K}/2\pm\mathbf{k},s}a_{\mathbf{K}/2\pm\mathbf{k},s}$ are Fermi number operators. Since (9) are not precisely Bose commutation rules it has been claimed [10], correctly, that the BCS operators do *not* stand, strictly speaking, for true bosons. Similarly to (8), the second-quantized creation and annihilation operators $c^{\dagger}_{\mathbf{K}}$ and $c_{\mathbf{K}}$ of 2h-CPs (see appendix) in their one-*hole* states of CMM wavevector $\mathbf{K}$ are postulated to be Bose operators and thus satisfy the relations

$$[c^{\dagger}_{\mathbf{K}}, c^{\dagger}_{\mathbf{K}'}] = 0, \qquad [c_{\mathbf{K}}, c_{\mathbf{K}'}] = 0, \qquad [c_{\mathbf{K}}, c^{\dagger}_{\mathbf{K}'}] = \delta_{\mathbf{K}',\mathbf{K}}. \tag{10}$$

It is precisely the inclusion of hole pairs along with the usual particle (here, electron) pairs that makes the present BF model a "*complete*" one.

By postulating both kinds of pairs one accounts for the factor of 2 in the smallest or elementary fluxon $\hbar/2e$ found in magnetic-flux quantization experiments in both conventional [35, 36] as well as in cuprate [37] superconductors. Here e is the *magnitude* of the charge of a single electron as, to our knowledge [44], these measurements do not determine the sign of the charge on the pair charge carriers. Although the vast majority of superconductors have *hole* charge carriers in the normal state [51], as determined empirically from *positive* Hall coefficients, there are London magnetic-moment measurements [52] establishing —at least according to one interpretation [53]— electron pairs in the condensates of superconductors as diverse as Pb (p-type above T_c), $BaPb_{0.8}Bi_{0.2}O_3$ (n-type) and $YBa_2Cu_3O_{7-\delta}$ (p-type), regardless of whether the single charge carriers in the normal state above T_c are electrons or holes. Related to this there is a wealth of experimental data [54–56] on the Hall-coefficient sign-reversal in various cuprates, but we shall not attempt to address in detail this phenomenon here.

We assume that creation and annihilation operators for particles of different kinds, i.e., of the fermions $(a^{\dagger}_{\mathbf{k},s}, a_{\mathbf{k},s})$ and of the 2e-CPs $(b^{\dagger}_{\mathbf{K}}, b_{\mathbf{K}})$, and of the 2h-CPs $(c^{\dagger}_{\mathbf{K}}, c_{\mathbf{K}})$ *commute* with each other. This is equivalent to assuming that the b and c operators describe *kinematically independent* composite bosons. Here we use for the unpaired-electron Fermi operators and for the 2e- and 2h-CP Bose operators the so-called "electronic" representation, see Appendix.

As to the *dynamical properties* of the many-fermion system postulated, they are described by the total Hamiltonian $H = H_0 + H_{int}$ equal to the sum of an unperturbed or zero-order Hamiltonian H_0 plus an interaction Hamiltonian H_{int}. Let us suppose further that the unperturbed Hamiltonian itself corresponds to a noninteracting BF mixture and is equal to sum of energy operators, respectively, of unpaired fermions, of 2e-CPs, and of 2h-CPs. Specifically

$$H_0 = \sum_{\mathbf{k}_1,s} \epsilon_{k_1} a^{\dagger}_{\mathbf{k}_1,s} a_{\mathbf{k}_1,s} + \sum_{\mathbf{K}} E_+(K) b^{\dagger}_{\mathbf{K}} b_{\mathbf{K}} - \sum_{\mathbf{K}} E_-(K) c^{\dagger}_{\mathbf{K}} c_{\mathbf{K}}. \tag{11}$$

As with the Fröhlich (or Dirac) interaction Hamiltonian for a BF mixture of phonons (or photons) and electrons, it is natural to suppose that the Hamiltonian for the specific interaction between unpaired fermions or holes and 2e-CPs and 2h-CPs has the form

$$\begin{aligned} H_{int} &= L^{-d/2} \sum_{\mathbf{k},\mathbf{K}} f_+(k) \left(a^{\dagger}_{\mathbf{k}+\frac{1}{2}\mathbf{K},\uparrow} a^{\dagger}_{-\mathbf{k}+\frac{1}{2}\mathbf{K},\downarrow} b_{\mathbf{K}} + a_{-\mathbf{k}+\frac{1}{2}\mathbf{K},\downarrow} a_{\mathbf{k}+\frac{1}{2}\mathbf{K},\uparrow} b^{\dagger}_{\mathbf{K}} \right) \\ &+ L^{-d/2} \sum_{\mathbf{k},\mathbf{K}} f_-(k) \left(a^{\dagger}_{\mathbf{k}+\frac{1}{2}\mathbf{K},\uparrow} a^{\dagger}_{-\mathbf{k}+\frac{1}{2}\mathbf{K},\uparrow} c^{\dagger}_{\mathbf{K}} + a_{-\mathbf{k}+\frac{1}{2}\mathbf{K},\downarrow} a_{\mathbf{k}+\frac{1}{2}\mathbf{K},\uparrow} c_{\mathbf{K}} \right) \end{aligned} \tag{12}$$

Operator terms in the first sum, which we characterize by some phenomenological function $f_+(k)$, describe the following elementary processes: a) of simultaneous annihilation of 2e-CPs with momentum wavevectors $\mathbf{K}$ and creation of two unpaired fermions with momenta wavevectors $\mathbf{k}_1 \equiv \mathbf{k} + \frac{1}{2}\mathbf{K}$ and $\mathbf{k}_2 \equiv -\mathbf{k} + \frac{1}{2}\mathbf{K}$—so that $\mathbf{k}_1 + \mathbf{k}_2 \equiv \mathbf{K}$ and $\mathbf{k} \equiv \frac{1}{2}(\mathbf{k}_1 - \mathbf{k}_2)$— and with the opposite spin projections; and b) of simultaneous creation of a two-fermion molecule with momentum wavevector $\mathbf{K}$ and annihilation of two unpaired fermions with momentum wavevectors $-\mathbf{k}+\frac{1}{2}\mathbf{K}$ and $\mathbf{k}+\frac{1}{2}\mathbf{K}$ and with opposite spin projections. These two elementary processes are analogous to the spontaneous decay (creation) of a two-fermion "molecule" into (from) two dissociated fermions. Operator terms in the second sum of (12), which we also characterize by some phenomenological function $f_{-}(k)$, describe the opposite elementary processes: a) simultaneous creation of two-hole molecules with momentum $\mathbf{K}$ and annhilation of two holes with momenta $-\mathbf{k}+\frac{1}{2}\mathbf{K}$ and $\mathbf{k}+\frac{1}{2}\mathbf{K}$ and with opposite spin projections. (This is an elementary process of creation of two-hole molecule during collision of two holes); 2) simultaneous annihilation of a 2h-CP with momentum $\mathbf{K}$ and creation of two fermions with momenta: $\mathbf{k}+\frac{1}{2}\mathbf{K}$ and $-\mathbf{k}+\frac{1}{2}\mathbf{K}$ and with opposite spin projections (This is an elementary process of spontaneous *decay* of two-holes molecules into two holes). In each of four elementary processes depicted in Fig. 1 total momentum and total spin projection of all particles which participate before and after an elementary process are conserved, as can be seen immediately from (12). Also, in each such elementary process the *total* number of fermions is conserved. The physical meaning of the phenomenological functions $f_{+}(k)$ and $f_{-}(k)$ is that they are momentum Fourier transforms of the internal wavefunctions of the *bound* 2e- and of 2h-CPs, respectively.

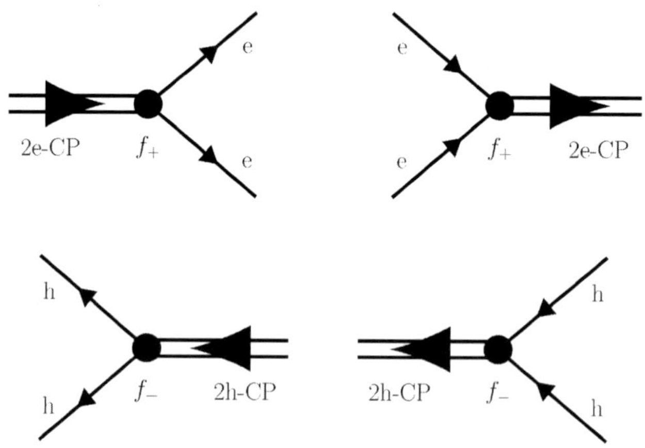

Figure 1. The BF Hamiltonian interaction (12) consists of four vertices each with two-fermion/one-boson creation-annihilation operators representing how unpaired electrons (+) and/or holes (-) bind to form 2e- or 2h-CPs, and disintegrate into two unpaired fermions.

Note that (11) plus (12) is quite different from the BCS Hamiltonian

$$H^{BCS} \equiv H_0^{BCS} + H_{int}^{BCS} = \sum_{\mathbf{k}_1, s_1} \epsilon_{k_1} a_{k_1,s_1}^{\dagger} a_{k_1,s_1} + \sum_{\mathbf{k}_1 \mathbf{l}_1} V_{\mathbf{k}_1,\mathbf{l}_1} a_{\mathbf{k}_1\uparrow}^{\dagger} a_{-\mathbf{k}_1\downarrow}^{\dagger} a_{-\mathbf{l}_1\downarrow} a_{\mathbf{l}_1\uparrow} \quad (13)$$

with, say, the *BCS model interaction*

$$V_{\mathbf{k}_1,\mathbf{l}_1} = \begin{cases} -V/L^d & \text{if } \mu - \hbar\omega_D < \epsilon_{k_1},\ \epsilon_{l_1} < \mu + \hbar\omega_D \\ 0 & \text{otherwise} \end{cases} \quad (14)$$

with $V \geqslant 0$. We shall consider the GBEC Hamiltonian (11) plus (12) with this latter interelectron interaction later on when writing down the precise conditions under which the GBEC theory contains the BCS theory as a special case.

3. GBEC Thermodynamics

We now discuss the *thermodynamic properties* of the postulated many-fermion system. Although *volume properties* of a many-fermion system are independent of the specific kind of macroscopic environment as in any statistical system, it is necessary from the very beginning to chose a definite type of environment in order to perform definite statistical-thermodynamics calculations.

We place the system in a macroscopic vessel of constant macroscopically large volume $\mathcal{V} \equiv L^d$. This vessel is in turn placed in an *environment* which we shall characterize by a constant given absolute temperature T, and by some given value of fermion chemical potential μ, i.e., let us assume that the walls of the vessel are *permeable* for heat but we maintain at all times a constant temperature T on them, and furthermore the walls are *permeable* for fermions so that one can maintain a definite constant value of the fermion chemical potential μ. Thus, let us allow our many fermion system to freely exchange with its environment both heat and fermions. Hence the *environment* is characterized by L^d, T and μ. In any statistical thermodynamic problem we must always suppose to have performed, at some step in the reasoning, a so-called *statistical limiting process* to an infinite system in which the volume of the system, its total energy, its total heat capacity, the total number of particles in the system, and in general any other so-called *extensive* thermodynamic quantity tends to infinity in such a way that volume density of energy, of thermodynamic potential, volume density of number of particles and, generally, the *volume density* of any extensive thermodynamic quantity remain *constant*. Such a limiting process we designate symbolically briefly as the "*TL*" (thermodynamic limit) $L^d \to \infty$, $N \to \infty$, assuming that L^d is the volume of the system and N its number of particles. For our many-fermion system we shall suppose that this statistical limiting process is performed. In particular, all sums over momenta must be replaced by the corresponding integrals over momenta, namely

$$\sum_{\mathbf{k}} \longrightarrow \left(\frac{L}{2\pi}\right)^d \int d^d k \equiv A_d \int dk\, k^{d-1}$$
$$\sum_{\mathbf{K}} \longrightarrow \left(\frac{L}{2\pi}\right)^d \int d^d K \equiv A_d \int dK\, K^{d-1} \tag{15}$$

for nonnegative k or K, where

$$A_d \equiv L^3/2\pi^2,\ L^2/2\pi \text{ and } L/\pi \tag{16}$$

for $d = 3, 2$ and 1, respectively. These replacements are an obvious consequence of the fact that the volume in 3D, or area in 2D, or length in 1D, of the elementary "cell" in momentum space is equal to $(2\pi/L)^d$ and tends to zero when $L^d \to \infty$.

Putting our many-fermion system in the environment L^d, T, μ then it will stay in some *thermodynamic equilibrium state*, i.e., in a state which is characterized by the values of thermodynamics parameters L^d, T, μ, where T is the environment absolute temperature. According to the *basic recipe of statistical thermodynamics* the thermodynamic (or "grand") potential $\Omega(T, L^d, \mu)$ corresponds to the environment L^d, T, μ. It is related to the system pressure P, internal energy E and entropy S by $\Omega = -PL^d = F - \mu N \equiv E - TS - \mu N$, where F is the Helmholtz free energy. It is defined as

$$\Omega(T, L^d, N, \mu) = -k_B T \ln \left(Tr[\exp\{-\beta(\hat{H} - \mu\hat{N})\}] \right) \tag{17}$$

where $\beta \equiv 1/k_B T$, Tr the "trace" of the diagonalized system, k_B the Boltzmann constant, and H is the *full* Hamiltonian of the system. Here $\hat{N}$ is the operator for the *total* number N of fermions in the system which for the BF system is just

$$\hat{N} = \sum_{\mathbf{k}_1,s} a^\dagger_{\mathbf{k}_1,s} a_{\mathbf{k}_1,s} + 2\sum_{\mathbf{K}} b^\dagger_{\mathbf{K}} b_{\mathbf{K}} - 2\sum_{\mathbf{K}} c^\dagger_{\mathbf{K}} c_{\mathbf{K}}. \tag{18}$$

The total number of fermions in the many-fermion system is the number of unpaired fermions, plus the number of fermions paired into 2e-CPs, *minus* the number of holes paired into 2h-CPs as these have opposite charge to the electrons. Thus, as a hole carries a charge of $+e$, if we multiply (18) throughout by the fermion charge $-e$ the result expresses the conservation of charge. From (11), (12) and (18) we immediately conclude that the full Hamiltonian $H = H_0 + H_{int}$ of our many-fermion system commutes with the operator $\hat{N}$ for the total number of fermions in the system.

The Gibbs identity for the thermodynamic potential Ω (17) is

$$d\Omega = -Pd\mathcal{V} - SdT - Nd\mu \tag{19}$$

where P is the pressure, S the entropy, and N the total number of fermions for that equilibrium state of the system when it has the constant volume L^d and is in contact with the environment held at constant temperature T and fermion chemical potential μ. As immediate consequences we have the familiar relations

$$P = -\left(\frac{\partial\Omega}{\partial\mathcal{V}}\right)_{T,\mu} \qquad S = -\left(\frac{\partial\Omega}{\partial T}\right)_{\mathcal{V},\mu} \qquad N = -\left(\frac{\partial\Omega}{\partial\mu}\right)_{T,\mathcal{V}} \tag{20}$$

where we assumed that the macroscopic environment of our many-fermion system is characterized by thermodynamic parameters L^d, T, and μ.

However, the characterization of the environment of a real thermodynamically equilibrium superconductor in experiments to measure, say, its specific heat, or critical magnetic field, when the superconductor carries no electric, characterizing the environment by the value of fermion chemical potential is unphysical. In such kind of experiments we have no free exchange of fermions between the superconductor and its environment. Physically, it is more correct to fix, in such traditional thermodynamic equilibrium conditions, not the value of fermion chemical potential μ of the environment but the value of the total number N of fermions in our many-fermion system. Thus, we shall do this in such a way that the system volume L^d, its temperature T, and its total number N of fermions are each held fixed. Thus we define our*environment* by L^d, T, N instead. In the TL the total electron number-density n remains constant, or

$$n = \frac{N}{L^d} \equiv const. \tag{21}$$

The main thermodynamic quantities for our many-fermion system in the environment characterized by T, L^d, n are then: the pressure $P = P(T, n)$, the entropy S per unit volume $s(T, n)$ and the constant-volume specific heat $C_\mathcal{V}$ per unit volume $c_\mathcal{V}(T, n)$. These quantities are clearly functions of the total number density n of electrons, and of the temperature T, namely

$$\frac{S}{\mathcal{V}} \equiv s(T,n), \qquad \frac{C_{\mathcal{V}}}{\mathcal{V}} \equiv c_{\mathcal{V}}(T,n). \tag{22}$$

Our many-fermion system, which consists of unpaired fermions, 2e-CPs and 2h-Cps, might in principle exhibit a peculiar physical property which one must correctly take into account. At some sufficiently small temperature T (perhaps zero) and at definite values of electron number-density n, the 2e- and 2h-CPs in the system can "condense" into their zero CMM states with $\mathbf{K} = 0$. In other words, macroscopically large numbers of 2e-CPs, N_0, or of 2h-CPs, M_0, or both, can occupy such states. At statistical thermodynamic limiting process $N \to \infty$, $\mathcal{V} \equiv L^d \to \infty$ the numbers of 2e-, N_0, and of 2h-, M_0, CPs condensed in the state $\mathbf{K} = 0$ tend to infinity ($N_0 \to \infty$, $M_0 \to \infty$) in such a way that the volume densities n_0 and m_0 of the condensed CPs remain finite, or

$$n_0 \equiv \frac{N_0}{L^d} = n_0(T,\mu) < \infty \qquad m_0 \equiv \frac{M_0}{L^d} = m_0(T,\mu) < \infty. \tag{23}$$

To ensure diagonalizability of the total Hamiltonian, following Bogoliubov [57] we allow for the possibility of such condensates of 2e-CPs and 2h-CPs in the state $\mathbf{K} = 0$ as follows. We shall go from initial or true to a reduced formulation of our mathematical quantum statistical problem as follows. Construct a new so-called *reduced* full Hamiltonian operator H' by replacing in the initial dynamical operator $\hat{H} - \mu\hat{N}$ the 2e-CP ($b_0^\dagger$, b_0) and 2h-CP ($c_0^\dagger$, c_0) creation and annihilation operators associated with the states $\mathbf{K} = 0$ by some as yet undefined ordinary or "b-", "c-" numbers $\sqrt{N_0}$ and $\sqrt{M_0}$, respectively. So we shall redefine our Hamiltonian from

$$H \equiv H_0 + H_{int} \qquad \text{to} \qquad H' \equiv H_0' + H_{int}'$$

with H_0' defined as

$$\begin{aligned} H_0' - \mu\hat{N} &\equiv [E_+(0) - 2\mu]N_0 + [2\mu - E_-(0)]M_0 + \sum_{\mathbf{k}} (\epsilon_k - \mu) a_{\mathbf{k},s}^\dagger a_{\mathbf{k}s} \\ &+ \sum_{\mathbf{K}\neq 0} [E_+(K) - 2\mu] b_{\mathbf{K}}^\dagger b_{\mathbf{K}} + \sum_{\mathbf{K}\neq 0} [2\mu - E_-(K)] c_{\mathbf{K}}^+ c_{\mathbf{K}} \\ &+ \sum_{\mathbf{k}} [\sqrt{n_0} f_+(k) + \sqrt{m_0} f_-(k)] \left(a_{\mathbf{k},\uparrow}^\dagger a_{-\mathbf{k}\downarrow}^\dagger + a_{-\mathbf{k}\downarrow} a_{\mathbf{k}\uparrow} \right) \end{aligned} \tag{24}$$

so that the new interaction Hamiltonian H_{int}' is now

$$\begin{aligned} H_{int}' &\equiv L^{-d/2} \sum_{\mathbf{k},\mathbf{K}\neq 0} f_+(k) \left(a_{\mathbf{k}+\frac{1}{2}\mathbf{K},\uparrow}^\dagger a_{-\mathbf{k}+\frac{1}{2}\mathbf{K},\downarrow}^\dagger b_{\mathbf{K}} + a_{-\mathbf{k}+\frac{1}{2}\mathbf{K},\downarrow} a_{\mathbf{k}+\frac{1}{2}\mathbf{K},\uparrow} b_{\mathbf{K}}^\dagger \right) \\ &+ L^{-d/2} \sum_{\mathbf{k},\mathbf{K}\neq 0} f_-(k) \left(a_{\mathbf{k}+\frac{1}{2}\mathbf{K},\uparrow}^\dagger a_{-\mathbf{k}+\frac{1}{2}\mathbf{K},\downarrow}^\dagger c_{\mathbf{K}}^\dagger + a_{-\mathbf{k}+\frac{1}{2}\mathbf{K},\downarrow} a_{\mathbf{k}+\frac{1}{2}\mathbf{K},\uparrow} c_{\mathbf{K}} \right) \end{aligned} \tag{25}$$

i.e., excluding the $\mathbf{K} = 0$ terms now included in (24). Note, that H_0' in (24) now depends *explicitly* on the numbers N_0 and M_0 which we have introduced as independent parameters.

Let us now insert the Hamiltonian H in the form $H_0' + H_{int}'$ in the basic statistical thermodynamic formula (17) and calculate the corresponding so-called modified thermodynamic potential

$$\Omega'(T, L^d, \mu, N_0, M_0) = -k_B T \ln\left(Tr[\exp\{-\beta(H_0' + H_{int}' - \mu\hat{N})\}]\right) \tag{26}$$

which now depends on the independent thermodynamic variables T, L^d, μ, N_0, M_0. It is related to the system pressure P', internal energy E' and entropy S' by

$$\Omega' = -P'L^d = F' - \mu N = E' - TS' - \mu N$$

where F' is the Helmholtz free energy. We fix the values of the numbers N_0 and M_0 in the following way. A thermodynamically stable state requires that for these values the Helmholtz free energy F', as taken for fixed T, L^d, μ, has its minimum value. In other words, let us demand that the numbers N_0 and M_0 satisfy

$$\frac{\partial F'}{\partial N_0} = 0 \qquad \frac{\partial F'}{\partial M_0} = 0 \qquad \text{and} \qquad \frac{\partial \Omega'}{\partial \mu} = -N \tag{27}$$

where the first partial derivatives are taken at constant T, L^d, μ, while the third derivative is the well-known result of mechanical statistics. Solving the system of eqs. (27) in respect to N_0 and M_0, we shall construct the following functions

$$N_0 = N_0(T, L^d, \mu) \qquad M_0 = M_0(T, L^d, \mu) \tag{28}$$

Solving the system of eqs. (27) we obtain specific values N_0 and M_0 for each set of values of independent thermodynamic parameters: T, L^d and μ, while the third equation in (27) ensures the charge conservation of the system.

Let us now substitute the functions (28) into the thermodynamic potential (26) and obtain some expression for the true thermodynamic potential of our many-fermion system in the environment characterized by T, L^d and μ. Explicitly, we have

$$\Omega(T, L^d, \mu) \equiv \Omega'[T, L^d, \mu, N_0(T, L^d, \mu), M_0(T, L^d, \mu)] \tag{29}$$

which depends now only on the variables T, L^d and μ. The equation (29) allows us to go from the independent variables T, μ to the independent variables T, n. To do so we use $-\partial\Omega'/\partial N = \mu$ in order to determine the function

$$\mu = \mu(n, T). \tag{30}$$

3.1. GBEC Eigenstates

Although the many-fermion system to be considered in this section is quite unrealistic, it is very simple and serves as a prelude in dealing with our full many-fermion system. Let us omit in the full dynamic operator (24)and (25) all interactions between unpaired fermions and condensed ($\mathbf{K} = \mathbf{0}$) as well as excited ($\mathbf{K} \neq 0$) bosonic CPs of both kinds, namely $2e$ and $2h$. Thus, the ideal boson-fermion model is defined by

$$\begin{aligned} H_0'' - \mu\hat{N} &= [E_+(0) - 2\mu]N_0 + [2\mu - E_-(0)]M_0 + \sum_{\mathbf{k},s}(\epsilon_k - \mu)a_{\mathbf{k},s}^\dagger a_{\mathbf{k},s} \\ &+ \sum_{\mathbf{K}\neq 0}[E_+(K) - 2\mu]\, b_{\mathbf{K}}^\dagger b_{\mathbf{K}} + \sum_{\mathbf{K}\neq 0}[2\mu - E_-(K)]c_{\mathbf{K}}^\dagger c_{\mathbf{K}}. \end{aligned} \tag{31}$$

This follows from (24) by putting $f_+(k) = f_-(k) = 0$. The operator H_0'' is clearly in diagonal form. Its *exact* eigenstates can be enumerated with the sets of occupation numbers $\{...n_{\mathbf{k},s}...N_{\mathbf{K}}...M_{\mathbf{K}}...\}$ where $n_{\mathbf{k},s}$ characterize one-particle electron states and take on two values 0 and 1, while one-boson occupation numbers $N_{\mathbf{K}}$ of 2e-CPs and $M_{\mathbf{K}}$ of 2h-CPs with momentum $\mathbf{K} \neq 0$, can take on values $0, 1, 2, \cdots\infty$. The exact eigenstates of the Hamiltonian H_0'' are then

$$|\, ...n_{\mathbf{k},s}...N_{\mathbf{K}}...M_{\mathbf{K}}...\rangle = \prod_{\mathbf{k},s}(a_{\mathbf{k},s}^\dagger)^{n_{\mathbf{k},s}} \prod_{\mathbf{K}\neq 0} \frac{1}{\sqrt{N_{\mathbf{K}}!}}\left(b_{\mathbf{K}}^\dagger\right)^{N_{\mathbf{K}}} \prod_{\mathbf{K}\neq 0} \frac{1}{\sqrt{M_{\mathbf{K}}!}}\left(c_{\mathbf{K}}^\dagger\right)^{M_{\mathbf{K}}} |0\rangle \tag{32}$$

where $|0\rangle$ is the vacuum state for fermions and simultaneously for 2e-CP and 2h-CP creation and annihilation operators. Specifically

$$a_{\mathbf{k},s}|0\rangle \equiv b_{\mathbf{K}}|0\rangle \equiv c_{\mathbf{K}}|0\rangle \equiv 0. \tag{33}$$

The exact eigenvalues of the unperturbed Hamiltonian H_0'' are thence

$$\begin{aligned} E_{\{...n_{\mathbf{k},s}...N_{\mathbf{K}}...M_{\mathbf{K}}...\}} &= [E_+(0) - 2\mu]N_0 + [2\mu - E_-(0)]M_0 + \sum_{\mathbf{k},s}(\epsilon_k - \mu)n_{\mathbf{k},s} \\ &+ \sum_{\mathbf{K}\neq 0}[E_+(K) - 2\mu]N_{\mathbf{K}} + \sum_{\mathbf{K}\neq 0}[2\mu - E_-(K)]M_{\mathbf{K}}. \end{aligned} \tag{34}$$

Let us now calculate for H_0'' the modified thermodynamic potential Ω_0'' with the help of its definition (26). In addition to definitions (4) and (6), we shall define

$$\begin{aligned} U_0 &\equiv [E_+(0) - 2\mu]N_0 + [2\mu - E_-(0)]M_0 \\ \xi_k &\equiv \epsilon_k - \mu \end{aligned} \tag{35}$$

One then immediately obtains for $Tr\left(\exp\left[-\beta(H_0^{''}-\mu\hat{N})\right]\right)$ the expression

$$\begin{aligned}
&\sum_{...n_{\mathbf{k},s}...N_{\mathbf{K}}...M_{\mathbf{K}}...} \langle ...n_{\mathbf{k},s}...N_{\mathbf{K}}...M_{\mathbf{K}}... \mid \exp\left[-\beta(H_0^{''}-\mu\hat{N})\right] \mid ...n_{\mathbf{k},s}...N_{\mathbf{K}}...M_{\mathbf{K}}...\rangle \\
&= \exp[-\beta U_0] \sum_{...n_{\mathbf{k},s}...N_{\mathbf{K}}...M_{\mathbf{K}}...} \exp[-\beta\xi_k n_{\mathbf{k},s}] \exp\left[-\beta\sum_{\mathbf{K}\neq 0}[E_+(\mathbf{K})-2\mu]\,N_{\mathbf{K}}\right] \\
&\times \exp\left[-\beta\sum_{\mathbf{K}\neq 0}[2\mu - E_-(K)]M_{\mathbf{K}}\right] \\
&= \exp[-\beta U_0]\prod_{\mathbf{k},s}\left[\sum_{n_{\mathbf{k},s}=0}^{1}\exp\left[-\beta\xi_k n_{\mathbf{k},s}\right]\right]\prod_{\mathbf{K}\neq 0}\left[\sum_{\mathbf{K}\neq 0}\exp\left[-\beta[E_+\mathbf{K}-2\mu]N_{\mathbf{K}}\right]\right] \\
&\times \prod_{\mathbf{K}\neq 0}\left[\sum_{\mathbf{K}\neq 0}\exp[-\beta[2\mu-E_-(K)]\,M_{\mathbf{K}}]\right] \\
&= \exp[-\beta U_0]\prod_{\mathbf{k},s}[1+\exp(-\beta\xi_k)]\prod_{\mathbf{K}\neq 0}[1-\exp[-\beta\,(E_+\mathbf{K}-2\mu)]]^{-1} \\
&\times \prod_{\mathbf{K}\neq 0}[1-\exp\{-\beta[2\mu-E_-(K)]\}]^{-1}. \qquad (36)
\end{aligned}$$

Since $\Omega_0^{''} = -k_BT\ln\left(Tr[\exp\{-\beta(H_0^{''}-\mu\hat{N})\}]\right)$ we have

$$\begin{aligned}
\Omega_0^{''} &= U_0 - 2k_BT\sum_{\mathbf{k}}\ln(1+\exp[-\beta\xi_k]) + k_BT\sum_{\mathbf{K}\neq 0}\ln\left(1-\exp[-\beta(E_+(K)-2\mu)]\right) \\
&+k_BT\sum_{\mathbf{K}\neq 0}\ln\left(1-\exp[-\beta(2\mu-E_-(K)])\right) \qquad (37)
\end{aligned}$$

Using (15) this quantity per unit volume $\mathcal{V}\equiv L^d$ becomes

$$\begin{aligned}
\Omega_0^{''}(T,\mu,n_0,m_0)/\mathcal{V} &= [E_+(0)-2\mu]n_0 + [2\mu - E_-(0)]m_0 \\
&- 2k_BT\left(\frac{A_d}{L^d}\right)\int_0^\infty k^{d-1}dk\ln(1+\exp[-\beta\xi_k]) \\
&+ k_BT\left(\frac{A_d}{L^d}\right)\int_{0+}^\infty K^{d-1}dK\ln\left(1-\exp[-\beta(E_+(K)-2\mu)]\right) \\
&+ k_BT\left(\frac{A_d}{L^d}\right)\int_{0+}^\infty K^{d-1}dK\ln\left(1-\exp[-\beta(2\mu-E_-(K))]\right). \qquad (38)
\end{aligned}$$

In general, for system pressure P, internal energy E and entropy S, one has $\Omega(T,L^d,\mu,N_0,M_0) = -PL^d = F-\mu N = E-TS-\mu N$ where F is the Helmholtz free energy. For the thermodynamic potential (38) we now require the conditions

$$\frac{\partial}{\partial n_0}\left(\frac{F_0''}{V}\right) \equiv [E_+(0) - 2\mu] = 0$$
$$\frac{\partial}{\partial m_0}\left(\frac{F_0''}{V}\right) \equiv [2\mu - E_-(0)] = 0 \tag{39}$$

while

$$n = -\frac{\partial}{\partial \mu}\left(\frac{\Omega_0''}{V}\right) \equiv 2n_0 - 2m_0 + 2\left(\frac{A_d}{L^d}\right)\int_0^\infty k^{d-1}dk\,[\exp[\beta\xi_k] + 1]^{-1}$$
$$+2\left(\frac{A_d}{L^d}\right)\int_0^\infty K^{d-1}dK\,[\exp[\beta(E_+(K) - 2\mu)] - 1]^{-1}$$
$$-2\left(\frac{A_d}{L^d}\right)\int_0^\infty K^{d-1}dK\,[\exp[\beta(2\mu - E_-(K))] - 1]^{-1} \tag{40}$$

where the first two relations are necessary conditions for thermodynamic stability while the third relation ensures constancy of the initial number of particles in the system. From (35) we have

$$\frac{\partial \xi}{\partial \mu} = -1 \qquad \frac{\partial[E_+(K) - 2\mu]}{\partial \mu} = -2 \qquad \frac{\partial[2\mu - E_-(K)]}{\partial \mu} = 2. \tag{41}$$

Note that the first equality of (40) in a real situation is satisfied only when the minimum of the thermodynamic potential is reached only when $n_0 \neq 0$ and $m_0 \neq 0$ are simultaneously satisfied, i.e., strictly within the region of available values $n_0 > 0$ and $m_0 > 0$. When the minimum is reached for $n_0 = 0$ and $m_0 \neq 0$, or for $n_0 \neq 0$ and $m_0 = 0$, or $n_0 = 0$ and $m_0 = 0$, i.e., on the *boundaries* of the region indicated, one of the first two equations of (40) or both must be omitted.

4. GBEC with Reduced BF Vertex Interaction

Let us omit in the full modified Hamiltonian H' (24), (25) the interaction operator H'_{int}, and approximately put $H' \simeq H'_0$. So we omit all the interaction between the fermions and excited 2e- and 2h-CPs, this interactions are studied in [58]. Thus we take into account only those BF interactions between the unpaired fermions or holes, and the 2e and 2h CPs in their respective condensates.

The modified dynamical operator H'_0 is given by the formula (24). This operator is a *quadratic form* in the fermion Fermi operators and consequently not difficult to determine *exactly* its eigenstates and eigenvalues. We need only transform the old creation and annihilation fermion operators $a^\dagger_{\mathbf{k},s}$ and $a_{\mathbf{k},s}$ to new creation and annihilation quasiparticle operators $\alpha^\dagger_{\mathbf{k},s}$ and $\alpha_{\mathbf{k},s}$ via the Bogoliubov-Valatin [57, 59] canonical (i.e., that preserve the form of Fermi anticommutation relations) u, v-transformation. Specifically

$$a_{\mathbf{k},s} = u_k\alpha_{\mathbf{k},s} + 2sv_k\alpha^\dagger_{-\mathbf{k},-s} \tag{42}$$

where $s = \pm\frac{1}{2}$ and

$$u_k^2 \equiv \tfrac{1}{2}(1 + \xi_k/E_k) \qquad v_k^2 \equiv \tfrac{1}{2}(1 - \xi_k/E_k) \tag{43}$$

which clearly satisfy $u_k^2 + v_k^2 = 1$ as well as

$$u_k v_k = \frac{\Delta_k}{2E_k}. \tag{44}$$

As before, the fermion energies are defined as

$$\xi_k \equiv \epsilon_k - \mu \qquad \text{and} \qquad E_k \equiv \sqrt{\xi_k^2 + \Delta_k^2} \tag{45}$$

with the fermion gap Δ_k given by

$$\Delta_k = \sqrt{n_0}f_+(k) + \sqrt{m_0}f_-(k). \tag{46}$$

On performing the u, v-transformation the operator $H_0' - \mu\hat{N}$ *exactly* becomes

$$\begin{aligned} H_0' - \mu\hat{N} &= U_{00} + \sum_{\mathbf{k},s} E_k \alpha_{\mathbf{k},s}^\dagger \alpha_{\mathbf{k},s} \\ &+ \sum_{\mathbf{K}\neq 0}[E_+(K) - 2\mu] b_{\mathbf{K}}^\dagger b_{\mathbf{K}} + \sum_{\mathbf{K}\neq 0}[2\mu - E_-(K)] c_{\mathbf{K}}^\dagger c_{\mathbf{K}}. \end{aligned} \tag{47}$$

This is because we now have *two*- instead of four-fermion operators as in the original BCS Hamiltonian so that no need arises to discard terms not in the form $\alpha_{\mathbf{k},s}^\dagger \alpha_{\mathbf{k},s}$. Here U_{00} is the ordinary number defined by

$$U_{00} \equiv [E_+(0) - 2\mu]N_0 + [2\mu - E_-(0)]M_0 + \sum_{\mathbf{k}}(\xi_k - E_k) \tag{48}$$

which differs from the previous number U_0 given in (35).

The exact eigenstates $|...\nu_{\mathbf{k},s}...N_{\mathbf{K}}...M_{\mathbf{K}}...\rangle$ and eigenvalues $E_{\{...\nu_{\mathbf{k},s}...N_{\mathbf{K}}...M_{\mathbf{K}}...\}}$ of the dynamical operator $H_0' - \mu\hat{N}$ (47) are then

$$|...\nu_{\mathbf{k},s}...N_{\mathbf{K}}...M_{\mathbf{K}}...\rangle = \prod_{\mathbf{k},s}(\alpha_{\mathbf{k},s}^\dagger)^{\nu_{\mathbf{k},s}} \prod_{\mathbf{K}\neq 0}\frac{1}{\sqrt{N_{\mathbf{K}}!}}\left(b_{\mathbf{K}}^\dagger\right)^{N_{\mathbf{K}}} \prod_{\mathbf{K}\neq 0}\frac{1}{\sqrt{M_{\mathbf{K}}!}}\left(c_{\mathbf{K}}^\dagger\right)^{M_{\mathbf{K}}} |\mathbf{0}\rangle \tag{49}$$

and

$$\begin{aligned} E_{[...\nu_{\mathbf{k},s}...N_{\mathbf{K}}...M_{\mathbf{K}}...]} &= U_{00} + \sum_{\mathbf{k},s} E_k \nu_{\mathbf{k},s} \\ &+ \sum_{\mathbf{K}\neq 0}[E_+(K) - 2\mu]N_{\mathbf{K}} + \sum_{\mathbf{K}\neq 0}[2\mu - E_-(K)]M_{\mathbf{K}}. \end{aligned} \tag{50}$$

Here $|\mathbf{0}\rangle$ is the vacuum state for new quasi-fermion creation $\alpha_{\mathbf{k},s}^\dagger$ and annihilation $\alpha_{\mathbf{k},s}$ operators, and simultaneously for the creation and annihilation 2e-CP and 2h-CP operators; it is different from the previous vacuum state $|0\rangle$ defined in (32). Namely,

$$\alpha_{\mathbf{k},s}|\mathbf{0}\rangle \equiv b_{\mathbf{K}}|\mathbf{0}\rangle \equiv c_{\mathbf{K}}|\mathbf{0}\rangle \equiv 0. \tag{51}$$

The symbol $[...\nu_{\mathbf{k},s}...N_{\mathbf{K}}...M_{\mathbf{K}}...]$ designates a *set* of values of the occupation numbers $\nu_{\mathbf{k},s}$ (which take on only the two values 0 or 1) of the *new* single quasi-fermion states $\mathbf{k}$, s, and of the values of the occupation numbers $N_{\mathbf{K}}$ and $M_{\mathbf{K}}$ of the single-boson states with $\mathbf{K} \neq 0$ of 2e-CPs and of 2h-CPs, respectively, which take on all values $0, 1, 2, \cdots, \infty$. Substituting (50) into (26) gives

$$\begin{aligned}
\Omega_0'(T, L^d, \mu, N_0, M_0) &= -k_B T \ln\left(Tr \exp\left[-\beta(H_0' - \mu\hat{N})\right]\right) \\
&= -k_B T \ln \sum_{...\nu_{\mathbf{k},s}...N_{\mathbf{K}}...M_{\mathbf{K}}...} \langle ...\nu_{\mathbf{k},s}...N_{\mathbf{K}}...M_{\mathbf{K}}... | \\
&\times \exp\left[-\beta(H_0' - \mu\hat{N})\right] | ...\nu_{\mathbf{k},s}...N_{\mathbf{K}}...M_{\mathbf{K}}...\rangle \\
&= [E_+(0) - 2\mu]N_0 + [2\mu - E_-(0)]M_0 \\
&+ \sum_{\mathbf{k}}(\xi_k - E_k) - 2k_B T \sum_{\mathbf{k}} \ln\left(1 + \exp[-\beta E_k]\right) \\
&+ k_B T \sum_{\mathbf{K}\neq 0} \ln\left(1 - \exp\left[-\beta(E_+(K) - 2\mu)\right]\right) \\
&+ k_B T \sum_{\mathbf{K}\neq 0} \ln\left(1 - \exp\left[-\beta(2\mu - E_-(K))\right]\right). \qquad (52)
\end{aligned}$$

Then the grand potential (29) becomes

$$\begin{aligned}
\Omega(T, L^d, \mu, N_0, M_0)/L^d &= [E_+(0) - 2\mu]n_0 + [2\mu - E_-(0)]m_0 \qquad (53) \\
&+ \int_0^\infty d\epsilon N(\epsilon)[\epsilon - \mu - E(\epsilon)] \\
&- 2k_B T \int_0^\infty d\epsilon N(\epsilon) \ln\left(1 + \exp[-\beta E(\epsilon)]\right) \\
&+ k_B T \int_{0+}^\infty d\varepsilon M(\varepsilon) \ln\left(1 - \exp[-\beta(E_+(0) + \varepsilon - 2\mu)]\right) \\
&+ k_B T \int_{0+}^\infty d\varepsilon M(\varepsilon) \ln\left(1 - \exp[-\beta(2\mu - E_-(0) + \varepsilon)]\right)
\end{aligned}$$

where

$$E(\epsilon) \equiv \sqrt{(\epsilon - \mu)^2 + \Delta^2(\epsilon)} \quad \text{where} \quad \Delta(\epsilon) \equiv \sqrt{n_0} f_+(\epsilon) + \sqrt{m_0} f_-(\epsilon) \qquad (54)$$

and the functions $f_+(\epsilon)$ and $f_-(\epsilon)$ are obtainable from the initial functions $f_+(k)$ and $f_-(k)$ with the help of replacement $k = \sqrt{2m\epsilon}/\hbar$.

One can rewrite these two expressions by using the first of the identities

$$2/(e^x + 1) \equiv 1 - \tanh\left(\tfrac{x}{2}\right) \qquad \text{and} \qquad 2/(e^x - 1) \equiv \coth\left(\tfrac{x}{2}\right) - 1. \qquad (55)$$

The first two transcendental system equations become

$$2\sqrt{n_0}[E_+(0) - 2\mu] = \int_0^\infty d\epsilon N(\epsilon) \frac{\Delta(\epsilon)\, f_+(\epsilon)}{E(\epsilon)} \tanh \tfrac{1}{2}\beta E(\epsilon) \qquad (56)$$

$$2\sqrt{m_0}[2\mu - E_-(0)] = \int_0^\infty d\epsilon N(\epsilon) \frac{\Delta(\epsilon)\, f_-(\epsilon)}{E(\epsilon)} \tanh \tfrac{1}{2}\beta E(\epsilon). \qquad (57)$$

On differentiating the modified thermodynamic potential $\omega_0' = \Omega_0'/V$ partially wrt μ for fixed values of n_0, m_0 and T one gets

$$\begin{aligned} -n &= \frac{\partial \omega_0'}{\partial \mu} = -\int_0^\infty d\epsilon N(\epsilon)\left(1+\frac{\partial E(\epsilon)}{\partial \mu}\right) + 2\int_0^\infty d\epsilon N(\epsilon)\frac{\exp[-\beta E(\epsilon)]}{1+\exp[-\beta E(\epsilon)]}\left(\frac{\partial E(\epsilon)}{\partial \mu}\right) \\ &- 2n_0 + \int_0^\infty d\varepsilon M(\varepsilon)\frac{\exp[-\beta(E_+(0)-2\mu+\varepsilon)]}{1-\exp[-\beta(E_+(0)-2\mu+\varepsilon)]}\left(\frac{\partial (E_+(0)-2\mu+\varepsilon]}{\partial \mu}\right) \\ &+ 2m_0 + \int_0^\infty d\varepsilon M(\varepsilon)\frac{\exp[-\beta(2\mu-E_-(0)+\varepsilon)]}{1-\exp[-\beta(2\mu-E_-(0)+\varepsilon)]}\left(\frac{\partial[2\mu-E_-(0)+\varepsilon]}{\partial \mu}\right). \end{aligned} \tag{58}$$

From (54) one has that

$$\frac{\partial E(\epsilon)}{\partial \mu} = -\frac{\epsilon-\mu}{E(\epsilon)}, \qquad \frac{\partial[E_+(0)-2\mu+\varepsilon]}{\partial \mu} = -2, \qquad \frac{\partial[E_+(0)-2\mu+\varepsilon]}{\partial \mu} = 2 \tag{59}$$

so that (58) becomes

$$\begin{aligned} n &= -\frac{\partial \omega_0'}{\partial \mu} = \int_0^\infty d\epsilon N(\epsilon)\left(1-\frac{\epsilon-\mu}{E(\epsilon)}\right) + 2\int_0^\infty d\epsilon N(\epsilon)\frac{\epsilon-\mu}{E(\epsilon)}\left(\exp[\beta E(\epsilon)\} + 1]^{-1}\right) \\ &+ 2n_0 + 2\int_{0+}^\infty d\varepsilon M(\varepsilon)\left(\exp[\beta(E_+(0)-2\mu+\varepsilon)]-1\right)^{-1} \\ &- 2m_0 - 2\int_{0+}^{+\infty} d\varepsilon M(\varepsilon)\left(\exp[\beta(2\mu-E_-(0)+\varepsilon)]-1\right)^{-1}. \end{aligned} \tag{60}$$

The third transcendental system equation hence becomes the *number equation*

$$n = 2n_B(T) - 2m_B(T) + n_f(T). \tag{61}$$

Here one defines

$$\begin{aligned} n_B(T) &\equiv n_0(T) + \int_{0+}^\infty d\varepsilon M(\varepsilon)[\exp \beta[E_+(0)-2\mu+\varepsilon]-1]^{-1} \\ &\equiv n_0(T) + n_{B+}(T) \end{aligned} \tag{62}$$

$$\begin{aligned} m_B(T) &\equiv m_0(T) + \int_{0+}^\infty d\varepsilon M(\varepsilon)[\exp \beta[2\mu-E_-(0)+\varepsilon]-1]^{-1} \\ &\equiv m_0(T) + m_{B+}(T) \end{aligned} \tag{63}$$

$$n_f(T) \equiv \int_0^\infty d\epsilon N(\epsilon)\left(1-\frac{\epsilon-\mu}{E(\epsilon)}\tanh\tfrac{1}{2}\beta E(\epsilon)\right) \tag{64}$$

in which $n_{B+}(T)$ and $m_{B+}(T)$ can be termed the total number-density of nonzero-CMM excited or "preformed" 2e-CPs and 2h-CPs, respectively, and $n_f(T)$ the number-density of unpaired fermions, all at a fixed temperature T. The system of *three* coupled integral equations (56), (57) and (61) consequently enables us to determine the *three* quantities $n_0 = n_0(T,n,\mu)$, $m_0 = m_0(T,n,\mu)$ and $\mu = \mu(T,n)$. The latter quantity can then be eliminated to give us $n_0 = n_0(T,n)$ and $m_0 = m_0(T,n)$. The thermodynamic properties can be obtained such as the entropy per unit volume $s(T,n)$ and the specific heat per unit volume $c_V(T,n)$.

Preformed bosonic objects *above* the condensate have apparently first been deduced to exist from gigahertz-frequency complex conductivity studies [60, 61] in $Bi_2Sr_2CaCu_2O_{8+\delta}$.

4.1. Symmetrical-Step Reduced BF Interaction

Instead of dealing with the phenomenological zero-CMM energies $E_+(0)$ and $E_-(0)$, respectively associated with the 2e- and 2h-CPs postulated to exist in the normal state, we define the two alternative *positive* energies

$$E_f \equiv \tfrac{1}{4}[E_+(0) + E_-(0)] \tag{65}$$

and

$$\delta\varepsilon \equiv \tfrac{1}{2}[E_+(0) - E_-(0)]. \tag{66}$$

The latter is positive provided only that $E_+(0) > E_-(0)$. Note that this is *not* satisfied by the original definition of CPs by Cooper based on the ideal Fermi gas "sea," where $E_+(0) < 0$ while $E_-(0) > 0$, but *is* satisfied by the Bethe-Salpeter treatments of Refs. [49] in 2D and [50] in 3D which are based on the BCS ground-state Fermi sea. Thus, the two energies E_f and $\delta\varepsilon$ replace $E_+(0)$ and $E_-(0)$ as our fundamental dynamical phenomenological parameters related to the 2e-and 2h-CPs. Both (65) and (66) imply that

$$E_\pm(0) \equiv 2E_f \pm \delta\varepsilon. \tag{67}$$

Consider now a reduced BF vertex interaction of symmetrical step-like form

$$f_+(\epsilon) = \begin{cases} f & \text{if} \quad E_f < \epsilon < E_f + \delta\varepsilon \\ 0 & \text{otherwise} \end{cases} \tag{68}$$

$$f_-(\epsilon) = \begin{cases} f & \text{if} \quad E_f - \delta\varepsilon < \epsilon < E_f \\ 0 & \text{otherwise.} \end{cases} \tag{69}$$

This particular form will allow making a one-to-one correspondence between the BF vertex (two-fermion) interaction and the familiar Cooper/BCS two-parameter model interelectron (four-fermion) interaction in the appropriate limits, as we see below, and through this association we obtain the BCS gap equation precisely. Both steps are perfectly adjacent to each other at the given characteristic energy E_f, have the same height f and the same width $\delta\varepsilon$.

The energy gap $\Delta(\epsilon)$ from (54) now becomes

$$\Delta(\epsilon) = \begin{cases} f\sqrt{n_0} & \text{if } E_f < \epsilon < E_f + \delta\varepsilon \\ f\sqrt{m_0} & \text{if } E_f - \delta\varepsilon < \epsilon < E_f \\ 0 & \text{otherwise,} \end{cases} \tag{70}$$

so that the electron energies become the gapped spectra

$$E(\epsilon) = \begin{cases} \sqrt{(\epsilon-\mu)^2 + f^2 n_0} & \text{if} \quad E_f < \epsilon < E_f + \delta\varepsilon \\ \sqrt{(\epsilon-\mu)^2 + f^2 m_0} & \text{if} \quad E_f - \delta\varepsilon < \epsilon < E_f \\ |\epsilon - \mu| & \text{otherwise.} \end{cases} \tag{71}$$

For a BF interaction of the form (52) and (53) our basic system of three transcendental equations (56) to (61) becomes, in view of (67),

$$[2E_f + \delta\varepsilon - 2\mu(T)] = \tfrac{1}{2} f^2 \int_{E_f}^{E_f+\delta\varepsilon} d\epsilon N(\epsilon) \frac{\tanh\left(\frac{1}{2}\beta\sqrt{[\epsilon-\mu(T)]^2 + f^2 n_0(T)}\right)}{\sqrt{[\epsilon-\mu(T)]^2 + f^2 n_0(T)}} \tag{i}$$

$$[2\mu(T) - 2E_f + \delta\varepsilon] = \tfrac{1}{2} f^2 \int_{E_f-\delta\varepsilon}^{E_f} d\epsilon N(\epsilon) \frac{\tanh\left(\frac{1}{2}\beta\sqrt{[\epsilon-\mu(T)]^2 + f^2 m_0(T)}\right)}{\sqrt{[\epsilon-\mu(T)]^2 + f^2 m_0(T)}} \tag{ii}$$

and

$$n = 2n_B(T) - 2m_B(T) + n_f(T) \tag{iii}$$

with $n \equiv N/L^d$ while the number density of the unpaired electrons is

$$n_f(T) \equiv \int_0^\infty d\epsilon N(\epsilon \left(1 - \frac{\epsilon-\mu}{E(\epsilon)} \tanh\left[\tfrac{1}{2}\beta E(\epsilon)\right]\right) \equiv 2\sum_{\mathbf{k}} v_{\mathbf{k}}^2(T). \tag{72}$$

One can also decompose the boson number densities as

$$n_B(T) \equiv n_0(T) + \int_{0+}^\infty d\varepsilon M(\varepsilon) \left(\exp[\beta(E_+(0) + \varepsilon - 2\mu) - 1]^{-1}\right) \equiv n_0(T) + n_{B+}(T)$$

$$m_B(T) \equiv m_0(T) + \int_{0+}^\infty d\varepsilon M(\varepsilon) \left(\exp[\beta(-E_-(0) + \varepsilon + 2\mu) - 1]^{-1}\right) \equiv m_0(T) + m_{B+}(T)$$

where $M(\varepsilon)$ is the bosonic density-of-states. Here $n_{B+}(T)$ and $m_{B+}(T)$ are the total number-density of nonzero-CMM preformed 2e-CPs and 2h-CPs, respectively. Note too that the $v_{\mathbf{k}}^2(T)$ are the finite-T generalization of the square of precisely the same $v_{\mathbf{k}}^2 \equiv 1 - u_{\mathbf{k}}^2$ coefficients that first appeared in the BCS trial wavefunction for singlet pairing given by

$$\prod_{\mathbf{k}} (u_{\mathbf{k}} + v_{\mathbf{k}} a_{\mathbf{k},\uparrow}^\dagger a_{-\mathbf{k},\downarrow}^\dagger)|0\rangle$$

where $|0\rangle$ is the BCS vacuum. The $u_{\mathbf{k}}$ and $v_{\mathbf{k}}$ also appear in the Bogoliubov-Valatin transformation equations (42) to (45).

5. Special Cases Subsumed in GBEC

In this section we describe how the GBEC formalism subsumes the five statistical theories of superconductivity as special cases. These five theories follow from GBEC in two principal branches: on the one hand the *symmetrical* case with equal proportions between 2e-CPs and 2h-CPs, and on the other hand the totally asymmetrical case where no 2h-CPs are present.

5.1. BCS-Bose (1967) Crossover

From the number equation (iii) above if one takes the ideal 50-50 proportion between 2e-CPs and 2h-CPs, i.e., $n_0(T) = m_0(T)$ one has

$$n_0(T) = m_0(T) = \Delta^2(T)/4f^2.$$

Adding (i) and (ii) gives

$$\delta\varepsilon = \tfrac{1}{2}f^2 \int_{E_f-\delta\varepsilon}^{E_f+\delta\varepsilon} d\epsilon N(\epsilon) \frac{\tanh\left(\frac{1}{2}\beta\sqrt{[\epsilon-\mu(T)]^2+\Delta^2(T)}\right)}{\sqrt{[\epsilon-\mu(T)]^2+\Delta^2(T)}}.$$

Furthermore, $n_{B+}(T) = m_{B+}(T)$ for all T implies from (i) and (ii) that

$$E_+(0) + \varepsilon_K - 2\mu = 2\mu - E_-(0) + \varepsilon_K$$

which in turn implies that

$$E_f = \mu.$$

Thus one arrives at

$$1 = \frac{f^2}{2\delta\varepsilon} N(0) \int_0^{\delta\varepsilon} d\xi \frac{1}{\sqrt{\xi^2+\Delta^2(T)}} \tanh\left(\tfrac{1}{2}\beta\sqrt{\xi^2+\Delta^2(T)}\right) \tag{73}$$

which is *precisely* the BCS gap equation [Ref. [10] Eq. (3.27)] in disguise, for all temperature T and all couplings, if and only if one makes the identification

$$\delta\varepsilon \equiv \hbar\omega_D \quad \text{and} \quad f^2/2\delta\varepsilon \equiv V.$$

This allows one to introduce the BCS dimensionless coupling parameter λ, namely

$$\frac{f^2}{2\hbar\omega_D} N(0) \equiv VN(0) \equiv \lambda \geqslant 0.$$

Thence, one is then left with

$$1 = \lambda \int_0^{\hbar\omega_D} d\xi \frac{1}{\sqrt{\xi^2+\Delta^2(T)}} \tanh\left(\tfrac{1}{2}\beta\sqrt{\xi^2+\Delta^2(T)}\right) \tag{74}$$

which is the more familiar form of the BCS gap equation.

In (iii) $n_B(T) = m_B(T)$ makes the first two terms cancel so that

$$n_f(T) \equiv \int_0^\infty d\epsilon N(\epsilon)\left(1 - \frac{\epsilon-\mu}{E(\epsilon)}\tanh\left[\tfrac{1}{2}\beta E(\epsilon)\right]\right) = n \equiv \frac{N}{L^d}. \tag{75}$$

Note that the two coupled equations (74) and (75) for $\Delta(T)$ and $\mu(T)$ define the "*BCS-Bose crossover* picture" first introduced by Friedel [62] et al. (1967). It was subsequently developed by Eagles [63] (1969); Leggett [64] (1980); Miyake [65] (1983); Nozières [66] (1985); Ranninger, Micnas & Robaszkiewicz [20] (1988); Randeria et al. [67] (1989); van der Marel [68] (1990); Bar-Yam [69] (1991); Drechsler & Zwerger [70] (1992); Haussmann [71] (1993); Pistolesi & Strinati [72] (1996) and many others.

One can has three different versions of the crossover [73], namely

A) $n = 2n_B(T) - 2m_B(T) + n_f(T)$.

B) If $n_0(T) = m_0(T)$ and $n_{B+}(T) = m_{B+}(T)$ implying $n = n_f(T)$.

C) Asymmetrical case with $m_B(T) = 0 \Rightarrow n = 2n_B(T) + n_f(T)$.

D) Asymmetrical case with $n_B(T) = 0 \Rightarrow n = 2m_B(T) + n_f(T)$.

In case (A) one has in addition the two-gap like equations (i) and (ii); in (B) one has equality between (i) and (ii) and in cases (C) and (D) one respectively has the gap-like equations (i) and (ii).

5.2. Condensation Energy at $T = 0$

It is a well-known fact that in conventional superconductivity a material becoming superconducting when it lowers its Helmholtz free energy [10, 74], i.e., $F_s(T) \leq F_n(T)$, for the absolute temperature $0 \leq T \leq T_c$, where $F_s(T)$ and $F_n(T)$ are the Helmholtz free energies for the superconducting and normal phases, respectively. Thus, $E_{cond}(T) \equiv F_s(T) - F_n(T)$ is the so-called condensation energy and is familiarly, albeit remarkably, several orders of magnitude smaller than typical atomic or even solid-state processes which are $\sim eVs$. For instance, the condensation energy for aluminium turns out to be as small as $4.42 \times 10^{-9} eV/atom$ [75] at $T = 0$; this is *eight* orders of magnitude smaller than the ionization energy $13.6\, eV$ of the hydrogen atom. However small, this tiny energy lowering apparently drives the second-order phase transition to the superconducting state. Thus, one would hope that any correct model of superconductivity should at least be consistent with this most striking empirical fact, among other phenomena. Here we show how the GBEC theory, when restricted to the BCS constraints of perfect two-electron/two-hole Cooper-pair (2e-/2h-CP) symmetry, gives analytic results for the condensation energy for all temperatures T below the critical T_c.

The ground-state ($T = 0$) condensation energy per unit volume of the superconducting relative to the normal phase according to the GBEC formalism, given (53), is

$$\frac{E_s - E_n}{L^d} = \frac{\Omega_s(T=0) - \Omega_n(T=0)}{L^d} \tag{76}$$

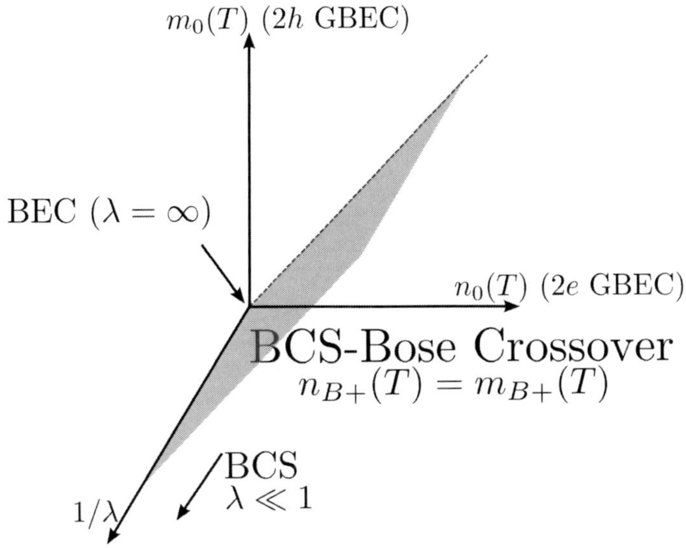

Figure 2. Parameter octant defined by the two condensate densities $n_0(T) \geq 0$ and $m_0(T) \geq 0$ as well as the (also non-negative) inverse $1/\lambda \geq 0$ of the interelectronic BCS dimensionless coupling $\lambda \geq 0$, and applicable in principle at all temperatures T. GBEC describes a ternary gas and applies in the entire octant. The BCS-Bose crossover theory applies only on the shaded plane defined by $n_0(T) = m_0(T)$ provided the additional restriction $n_{B+}(T) = m_{B+}(T)$ is imposed whereby the total number of 2e (two-electron) noncondensate CPs equals that of 2h (two-hole) CPs. BCS theory is valid *only* along the forefront of the shaded plane where $\lambda \ll 1$ of the shaded BCS-Bose crossover plane since weak coupling then justifies $\mu \simeq E_F$ as assumed by BCS which reduces the number equation to a triviality leaving only the gap equation. This plot was inspired in the celebrated Bronstein cube.

since for any T the Helmholtz free energy $F \equiv E - TS = \Omega + \mu N$, with S the entropy, and μ is the same for either superconducting s or normal n phases with internal energies E_s and E_n, respectively. In the *normal* phase $n_0(T) = 0$, $m_0(T) = 0$ so that $\Delta(T) = 0$ for all $T \geq 0$, and (53) reduces to

$$\frac{\Omega_n(T=0)}{L^d} = \int_0^\infty d\epsilon N(\epsilon)(\epsilon - \mu - |\epsilon - \mu|) = 2\int_0^\mu d\epsilon N(\epsilon)(\epsilon - \mu) = 2\int_{-\mu}^0 d\xi N(\xi)\xi. \tag{77}$$

For the superconducting phase, and when $n_0(T) = m_0(T)$ and $n_B(T) = m_B(T)$ hold, i.e., in the crossover scenario, one deduces from (67) and (53) that $\mu = E_f$. Putting $\Delta(T=0) \equiv \Delta$ in (53) as well as $\delta\varepsilon \equiv \hbar\omega_D$, while using (67), gives

$$\begin{aligned} \frac{\Omega_s(T=0)}{L^d} &= 2\hbar\omega_D\, n_0(0) + \int_{-\mu}^\infty d\xi N(\xi)\left(\xi - \sqrt{\xi^2 + \Delta^2}\right) \\ &= 2\hbar\omega_D\, n_0(0) + 2\int_{-\mu}^{-\hbar\omega_D} d\xi N(\xi)\xi - 2\int_0^{\hbar\omega_D} d\xi N(\xi)\sqrt{\xi^2 + \Delta^2} \end{aligned} \tag{78}$$

The first factor of 2 in the last line comes precisely from the condition $n_0(T) = m_0(T)$ while the last two factors of 2 arise from the condition that according to (68) and (69) the magnitudes of $f_+(\epsilon)$ and $f_-(\epsilon)$ are the *same* and equal f. Subtracting (77) from (78) and putting $N(\xi) \cong N(0)$, the density of electronic states at the Fermi surface—designated before as $N(E_F)$—yields by standard integrations

$$\begin{aligned}\frac{E_s - E_n}{L^d} &= 2\hbar\omega_D\, n_0(0) + 2N(0)\int_0^{\hbar\omega_D} d\xi\left(\xi - \sqrt{\xi^2+\Delta^2}\right) \\ &= 2\hbar\omega_D\, n_0(0) + N(0)\Big[(\hbar\omega_D)^2 - \hbar\omega_D\sqrt{(\hbar\omega_D)^2+\Delta^2} \\ &+ \Delta^2 \ln\frac{\Delta}{\hbar\omega_D + \sqrt{(\hbar\omega_D)^2+\Delta^2}}\Big] \qquad \text{(GBEC)} \qquad (79)\end{aligned}$$

exactly. Using the expression that follows from (73) for $T = 0$ gives Eq. (2.40) of Ref. [10], namely

$$\Delta = \frac{\hbar\omega_D}{\sinh(1/\lambda)} \qquad (80)$$

where λ is related to the BF interaction parameter f of the GBEC formalism through

$$\lambda \equiv VN(0) = \frac{f^2 N(0)}{2\hbar\omega_D}.$$

This makes the first term on the rhs of (79) exactly equal to $\Delta^2 N(0)/\lambda$ which in turn can be shown to cancel exactly against the log term if one recalls the hyperbolic-function identity $\sinh^2 x + 1 \equiv \cosh^2 x$. Thus, the GBEC formalism condensation energy (79) is identical for *any* coupling to that of BCS theory, Eq. (2.42) of Ref. [10], namely

$$\begin{aligned}\frac{E_s - E_n}{L^d} &= N(0)(\hbar\omega_D)^2\left[1 - \sqrt{1+(\Delta/\hbar\omega_D)^2}\right] \qquad \text{(BCS)} \qquad (81) \\ &\xrightarrow[\lambda\to 0]{} -\frac{1}{2}N(0)\Delta^2\left[1 - \frac{1}{4}\left(\frac{\Delta}{\hbar\omega_D}\right)^2 + O\left(\frac{\Delta}{\hbar\omega_D}\right)^4\right].\end{aligned}$$

This energy, associated with the expectation value of the BCS trial wavefunction gives a rigorous upper bound to the exact ground-state energy of the BCS Hamiltonian. Empirically, for niobium (Nb, bcc, $T_c \simeq 9.3K$, critical magnetic field $H_c \simeq 160$ kA/m) the condensation energy to be compared with the BCS result (81) works out to be just 2×10^{-6}eV/atom [75]. The equivalence of (79) and (81) seems to suggest that, as in the GBEC formalism, there are no pair-pair interactions in the BCS theory either, as is evident from Hamiltonians (11), (12) and (14). See appendix B for the condensation energy at $T \geq 0$.

5.3. BCS (1957) Theory

Ordinary BCS theory is characterized by a single equation, namely, the gap equation for all coupling $\lambda \equiv VN(0)$ and temperatures T. The auxiliary (number) equation is replaced by the assumption that $\mu \simeq E_F$. If $\xi \equiv \epsilon - \mu$, and $\hbar\omega_D << \mu$,

$$1 \simeq VN(\mu) \int_0^{\hbar\omega_D} d\xi \frac{1}{\sqrt{\xi^2 + \Delta(T)^2}} \tanh\left[\tfrac{1}{2}\beta\sqrt{\xi^2 + \Delta(T)^2}\right]$$

is BCS gap equation (3.27) in 1957 BCS paper [10].

For *weak coupling* $\Delta \to 0$ gives $\mu \simeq E_F$ as assumed in ordinary BCS theory, since

$$n = \int_0^\infty d\epsilon N(\epsilon)\left(1 - \frac{\epsilon - \mu}{E(\epsilon)} \tanh\left[\tfrac{1}{2}\beta E(\epsilon)\right]\right) \equiv n_f(T)$$

$$n_f(T) \underset{\Delta \to 0}{\longrightarrow} \int_0^\infty d\epsilon N(\epsilon)\left(1 - \tanh \tfrac{1}{2}\beta[\epsilon - \mu]\right) \equiv 2\int_0^\infty d\epsilon N(\epsilon)\left(\exp[\beta(\epsilon - \mu(T))] + 1\right)^{-1}$$

given that $|x|^{-1}\tanh|x| \equiv x^{-1}\tanh x$ for all $-\infty < x < \infty$ and $1 - \tanh\frac{1}{2}x \equiv 2/(\exp^x + 1)$, and similarly $\coth\frac{1}{2}(x-1) \equiv 2/(e^x - 1)$ so that

$$n = 2\int_0^{\mu(0)} d\epsilon N(\epsilon) = 2\left(\frac{m^{3/2}}{2^{1/2}\pi^2\hbar^3}\right)\int_0^{E_F} d\epsilon\sqrt{\epsilon} = \frac{2\sqrt{2}m^{3/2}E_F^{3/2}}{3\pi^2\hbar^3} \quad \Rightarrow$$

$$E_F \equiv \frac{\hbar^2\left(3\pi^2 n\right)^{2/3}}{2m}.$$

Thus, the BCS gap equation in GBEC can be viewed as the condition for thermodynamic phase equilibrium between two *pure* BEC phases of 2e- and 2h-CPs, if coupling is weak. The energy gap equation at $T = 0$ readily integrated exactly gives

$$\Delta(0) = \frac{\hbar\omega_D}{\sinh(1/\lambda)} \underset{\lambda \to 0}{\longrightarrow} 2\hbar\omega_D \exp\left(-1/\lambda\right),$$

where dimensionless coupling parameter $\lambda \equiv f^2 N(E_F)/2\hbar\omega_D \equiv VN(E_F)$.

Finally, $\Delta(T_c) = 0$ leads to

$$\frac{1}{\lambda} = \int_0^{\hbar\omega_D/2k_BT_c} dx\, \frac{\tanh x}{x}$$

or the weak-coupling BCS T_c formula

$$k_BT_c \underset{\lambda \to 0}{\longrightarrow} (2e^\gamma/\pi)\hbar\omega_D \exp\left(-1/\lambda\right) \simeq 1.134\hbar\omega_D \exp\left(-1/\lambda\right)$$

if $\Theta_D/2T_c >> 1$, where $\Theta_D \equiv \hbar\omega_D/k_B$ is the Debye temperature [76]. This implies the gap-to-T_c ratio

$$\frac{2\Delta(0)}{k_BT_c} = 2\pi\exp(-\gamma) \simeq 3.53$$

where the Euler constant $\gamma \simeq 0.5772$.

5.4. Friedberg-T. D. Lee (1989) BEC Theory

We take now the second branch of GBEC, the complete asymmetrical case with only 2e pairs, i.e., if set $m_B(T) \equiv 0$ and $m_0(T) = 0$ implying $f_-(\epsilon) \equiv 0$, therefore (ii) drops out from set $\{(i), (ii), (iii)\}$. If one assumes the interaction

$$f_+(\epsilon) = \begin{cases} f & \text{if } E_f - \delta\varepsilon < \epsilon < E_f + \delta\varepsilon \\ 0 & \text{otherwise} \end{cases}$$

and pick $E_f = E_+(0)/2$. Then (i) becomes

$$4[E_f - \mu] = f^2 \int_{E_f-\delta\varepsilon}^{E_f+\delta\varepsilon} d\epsilon \frac{N(\epsilon)}{\sqrt{(\epsilon-\mu)^2 + f^2 n_0}} \tanh\left(\tfrac{1}{2}\beta\sqrt{(\epsilon-\mu)^2 + f^2 n_0}\right) \qquad (82)$$

Also, (iii) reduces to $n = 2n_B(T) + n_f(T)$, or

$$\begin{aligned} n &= 2n_0 + 2\int_{0^+}^{\infty} d\varepsilon M(\varepsilon)\left(\exp[\beta(E_+(K) - 2\mu)] - 1\right)^{-1} \\ &+ \int_0^{\infty} d\epsilon N(\epsilon)\left(1 - \frac{\epsilon-\mu}{E(\epsilon)}\tanh\left[\tfrac{1}{2}\beta E(\epsilon)\right]\right) \end{aligned} \qquad (83)$$

where $E_+(K) - 2\mu \equiv E_+(0) + \varepsilon - 2\mu = 2(E_f - \mu) + \varepsilon$ and

$$E(\epsilon) = \begin{cases} \sqrt{(\epsilon-\mu)^2 + f^2 n_0} & \text{if } E_f - \delta\varepsilon < \epsilon < E_f + \delta\varepsilon \\ |\epsilon - \mu| & \text{otherwise.} \end{cases}$$

Eqs. (82) and (83) are *precisely* (4.3) and (4.4) of Friedberg and Lee (1989) with the following notational changes (GBEC ⇒ F-T.D. Lee)

$$E_+(0) \Rightarrow 2\nu_0 \qquad L^3 \Rightarrow \Omega \qquad f \Rightarrow g \qquad n_0 \Rightarrow |B|^2 \qquad n \Rightarrow \rho.$$

5.5. Ideal GBEC

If one turns "off" the interaction in GBEC, i.e., $f = 0$, this formalism becomes an ideal boson-fermion model (IBFM), see Fig.3. The IBFM is completely described in d dimensions by the Hamiltonian H_0 defined by (11), but without interactions. One can construct its associated grand potential as

$$\Omega_0''(T, L^d, \mu, N_0, M_0) = -k_B T \ln\left[Tr\ \exp\{-\beta(H_0'' - \mu\hat{N})\}\right] \qquad (84)$$

where H_0'' is (31), with N_0 and M_0 as before the number of zero-CMM 2e- and 2h-CPs, respectively. Omitted double prime, one gets

$$\begin{aligned}\Omega_0(T, L^d, \mu, N_0, M_0) &= [E_+(0) - 2\mu]N_0 + [2\mu - E_-(0)]M_0 \\ &- 2k_B T B_d L^d \int_0^\infty k^{d-1} dk \ln\left(1 + \exp[-\beta(\epsilon_k - \mu)]\right) \\ &+ k_B T B_d L^d \int_{0+}^\infty K^{d-1} dK \ln\left(1 - \exp[-\beta(E_+(K) - 2\mu)]\right) \\ &+ k_B T B_d L^d \int_{0+}^\infty K^{d-1} dK \ln\left(1 - \exp[-\beta(2\mu - E_-(K))]\right) \quad (85)\end{aligned}$$

where $B_d = 1/\pi, 1/2\pi$ and $1/2\pi^2$ for $d = 1, 2$ and 3, respectively. An equilibrium thermodynamic state makes $\Omega_0(T, L^d, \mu, N_0, M_0)$ stationary with respect to N_0 and to M_0 and requires that the number density of electrons $n \equiv N/L^d$ remains constant. Thus, one imposes that

$$\frac{\partial F}{\partial N_0} = 0 \qquad \frac{\partial F}{\partial M_0} = 0 \qquad \text{and} \qquad -\frac{\partial \Omega_0}{\partial \mu} = N. \quad (86)$$

where $F = (L^d, T, \mu, N_0, M_0) = -k_B T \ln\left(Tr \exp[-\beta\Omega]\right)$ is the Helmholtz free energy. This leads to the three relations

$$[E_+(0) - 2\mu] = 0 \qquad [2\mu - E_-(0)] = 0 \qquad \text{and} \qquad n_f(T) + 2n_B(T) - 2m_B(T) = n$$

with the latter being the *"number equation"* ensuring charge conservation in the ternary mixture. In the last relation

$$n_B(T) \equiv n_0(T) + B_d \int_{0+}^\infty dK K^{d-1} \left(\exp[\beta(E_+(0) - 2\mu + \varepsilon_K)] - 1\right)^{-1} \quad (87)$$

$$m_B(T) \equiv m_0(T) + B_d \int_{0+}^\infty dK K^{d-1} \left(\exp[\beta(2\mu - E_-(0) + \varepsilon_K)] - 1\right)^{-1} \quad (88)$$

while

$$n_f(T) \equiv 2B_d \int_{0+}^\infty dk\, k^{d-1} \left(\exp[\beta(\epsilon_k - \mu)] + 1\right)^{-1}. \quad (89)$$

This last expression can be interpreted as the number density of *unpaired* electrons at any T and $\epsilon_k = \hbar^2 k^2/2m$ is the electron energy.

5.6. Ideal BF Model of Casas et al. (1998-2002)

The ideal BF model of Casas et al. [32–34] is simple statistical model treating CPs as non-interacting bosons in thermal and chemical equilibrium with unpaired fermions is proposed. The model gives rise to a boson number that is strongly coupling and temperature-dependent. This model naturally suggest a more convenient definition of the boson chemical potential whereby the generate Fermi region of positive fermion chemical potential can be accessed unlike previous treatments. The authors provide support for a widespread conjecture (or 'paradigm') that superconductivity in general is a Bose-Einstein condensation of the charged Cooper pairs.

For *noninteracting* or ideal BF model, or IBFM, i.e., for $f = 0$, $E(\epsilon) = |\epsilon - \mu|$ and (82) implies $E_f = \mu$; so $E_+(K) - 2\mu = 2(E_f - \mu) + \varepsilon = \varepsilon$ and hence only (iii) is left with $\Delta = 0$, namely

$$n = 2\,n_0 + 2\int\limits_{0+}^{\infty} d\varepsilon M(\varepsilon)\left(\exp[\beta\varepsilon] - 1\right)^{-1} + 2\int\limits_{0}^{\infty} d\epsilon N(\epsilon)\left(\exp[\beta(\epsilon - E_f)] + 1\right)^{-1}$$

or more briefly

$$2n_B(T) + n_f^0(T) = n.$$

Not that in 2D

$$\begin{aligned} n_f(T) \underset{f\to 0}{\longrightarrow} n_f^0(T) &\equiv 2\left(\frac{m}{2\pi\hbar^2}\right)\int\limits_{0}^{\infty} d\epsilon\left(\exp[\beta(\epsilon - E_f)] + 1\right)^{-1} \\ &= \left(\frac{m}{\pi\hbar^2}\right)\left(E_f + \beta^{-1}\ln\{1 + \exp[-\beta E_f]\}\right) \\ &\equiv n_f\left(1 + \left(\frac{T}{T_f}\right)\ln\{1 + \exp[-T_f/T]\}\right) \geqslant n_f \end{aligned}$$

exactly, where $n_f \equiv mE_f/\pi\hbar^2$. Clearly, this model might be useful in locating the superconducting singularity in temperature T_c as the system is cooled down, but it cannot describe the superconducting phase as such since the gap is zero.

5.7. Original BEC Theory (1925)

Critical temperature T_c —defined as that value of T below which $n_0(T)$ *just ceases to be zero* as system is cooled and obtained by putting $n_0(T_c) = 0$, or

$$\begin{aligned} 2n_B(T_c) &\equiv 0 + 2\int\limits_{0+}^{\infty} d\varepsilon M(\varepsilon)\left(\exp[\beta_c\varepsilon] - 1\right)^{-1} \\ &= n - 2\int\limits_{0}^{\infty} d\epsilon N(\epsilon)\left(\exp[\beta_c(\epsilon - E_f)] + 1\right)^{-1} \equiv n - n_f^0(T_c) \end{aligned}$$

with $\beta_c \equiv 1/k_B T_c$ and $n_f^0(T_c)$ is the *number density of unpaired electrons* in the IBFM at T_c. If $M(\varepsilon) \equiv (2m^{3/2}/\pi^2\hbar^3)\sqrt{\varepsilon}$ we get, since

$$\int\limits_{0}^{\infty} dx\frac{\sqrt{x}}{e^x - 1} = \Gamma(3/2)\zeta(3/2)$$

then

$$T_c \simeq \left(\frac{2\pi}{(2.612)^{2/3}}\right)\left(\frac{\hbar^2}{2mk_B}\right)\left(\frac{n-n_f^0(T_c)}{2}\right)^{2/3} \equiv \left(\frac{2\pi}{(2.612)^{2/3}}\right)\left(\frac{\hbar^2}{2mk_B}\right)n_B(T_c)^{2/3}$$
$$\simeq 3.31\left(\frac{\hbar^2}{2mk_B}\right)n_B(T_c)^{2/3}.$$

Since there are still a very tiny fraction of unpaired electrons in the BF gas the expression here is an *implicit* formula for the ideal boson gas (IBG) with boson particle density $n_B(T_c) \equiv \frac{1}{2}[n - n_f^0(T_c)]$; otherwise, it is of precisely the same *form* as the ordinary BEC T_c-formula. Note that

$$n_f^0(T) \simeq n_f\left(1 + \frac{T}{T_f}\exp[-T/T_f] + \cdots\right) \geq n_f$$

This can also be thought of as an *extreme strong-coupling* regime whereby $n_f^0(T)$ vanishes so that $n_B(T_c) = n/2$ and one recovers the familiar limit

$$\frac{T_c}{T_F} \simeq 0.218.$$

5.8. GBEC Phases

The equations system $\{(i),(ii),(iii)\}$ can be solved for T_c/T_F as a function of n/n_f. One has a different phase boundary, i.e., with a pure phase of 2h-GBEC, another one for 2e-GBEC and a third mixed phase with arbitrary proportions between 2e-CPs and 2h-CPs, in this mixed phase appears the BCS-Bose crossover with an ideal 50-50 symmetry between 2e-CPs and 2h-CPs. This diagram phases is plotted in Fig. 4 using the BCS dimensionless coupling constant $\lambda = 1/2$ and $\hbar\omega_D/E_F = 10^{-3}$, where $\hbar\omega_D$ is the Debye energy of the lattice.

Here we reproduce the BCS value from the BCS T_c formula

$$T_c/T_F \simeq 1.134(\hbar\omega_D/E_F)\exp(-1/\lambda)$$

which is $\simeq 7.64\times10^{-6}$ (large red dot in Fig. 4); it lies well within the range $T_c/T_F \lesssim 10^{-3}$ empirically found for conventional, elemental SCs. Empirical data [77] for "exotic" SCs, however, fall within the range $T_c/T_F \simeq 0.01 - 0.05$. Thus, moderate departures from perfect 2e/2h-CP symmetry enable the ideal GBEC formalism to access, in sharp contrast to BCS theory, empirical T_c values for exotic SCs *without abandoning electron-phonon dynamics*.

Finally, for intermediate values of μ, namely for $E_-(0) < 2\mu < E_+(0)$, neither the first nor second equations of (86) are satisfied so that $n_0(T) = m_0(T) = 0$ for all T; this implies no condensed phases whatsoever. Thus, the ideal GBEC (characterized by zero BF coupling f) *contains no mixed phase* in contrast to the GBEC where $f \neq 0$ (Ref. [27] for a quadratic dispersion). This case will be treated elsewhere with the correct linear dispersion.

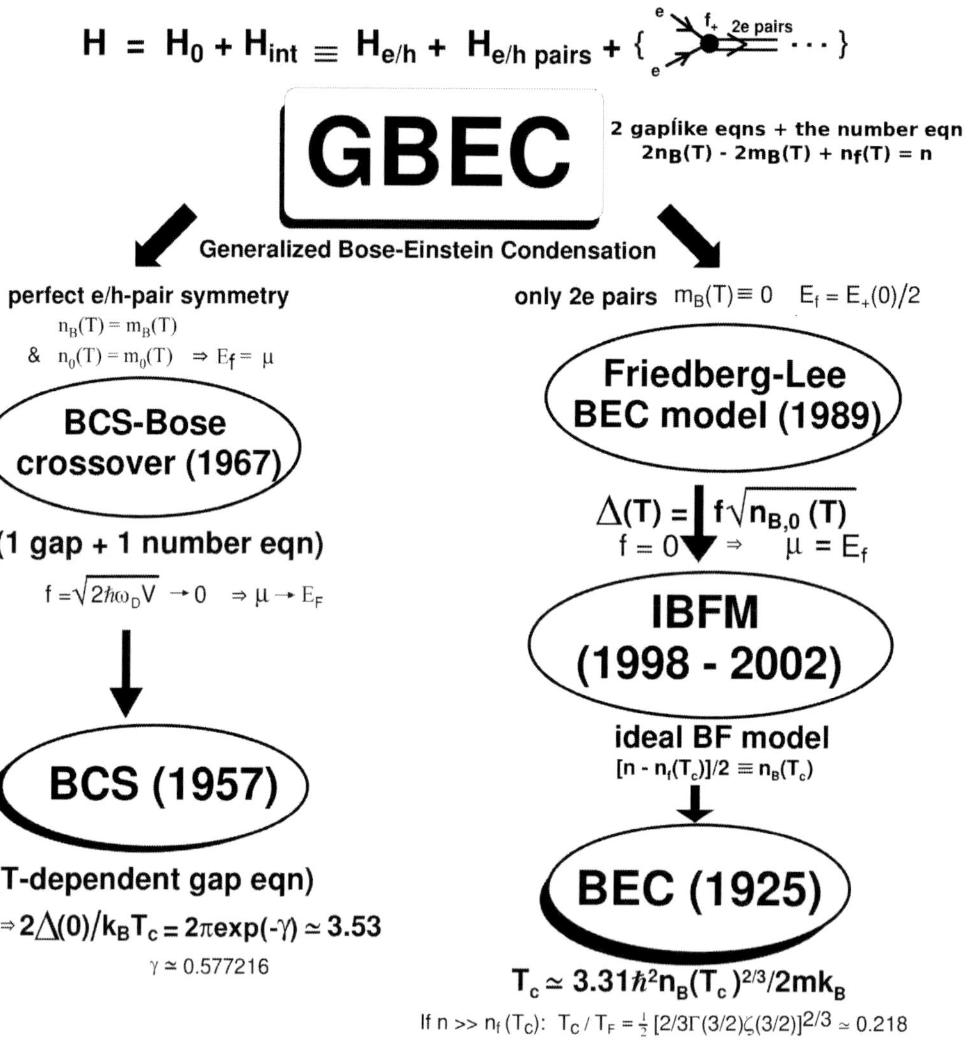

Figure 3. Flowchart illustrating conditions under which the GBEC formalism reduces to, or subsumes, all five statistical theories of superconductivity (ovals). Here ideal boson-fermion model (IBFM) is the *binary* case of GBEC corresponding to the unperturbed Hamiltonian (11) only. Here $\zeta(3/2) \simeq 2.612$ is the Zeta function of order 3/2.

6. Conclusions

The new generalized BEC (GBEC) theory includes as limiting cases the following theories: *i)* BCS, *ii)* BCS-Bose "*crossover*," when the BE condensate consists of equal numbers of electron- and hole-pairs. It also contains: *iii)* the Friedberg-T.D. Lee BEC model, *iv)* the "ideal boson-fermion model" (IBFM), and *v)* ordinary BEC theory when there are no unpaired fermions at all. The BCS condensate is precisely a BE condensate of equal numbers of 2e/2h-pairs with weak coupling. Without abandoning electron-phonon dynamics the GBEC theory leads to 2-to-3 order-of-magnitude higher T_c's –including room-temperature

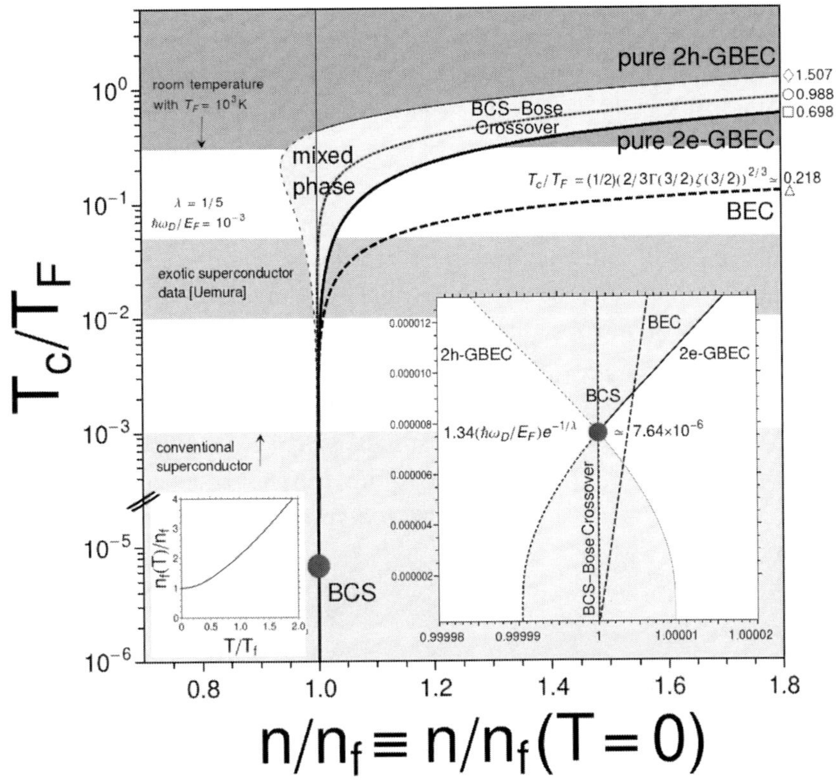

Figure 4. Dimensionless T_c/T_F versus n/n_f for pure GBEC phases 2h-/2e-CPs and the ordinary BEC in 3D (thick dashed curve), extrapolating for $n_f \to 0$ to the familiar limit 0.218. Compared are three bands encompassing conventional, exotic [77] and room-temperature empirical T_c values. Larger Inset is a blow up around $n/n_f = 1$ where the intersection reproduces the BCS value $T_c/T_F = 7.64 \times 10^{-6}$ given by the BCS T_c weak-coupling formula $k_B T_c \simeq 1.134\hbar\omega_D \exp(1/\lambda)$ for $\lambda = 1/5$ and $\hbar\omega_D = 10^{-3}E_F$, where $\hbar\omega_D$ is the Debye energy of the lattice. Red dot (online) marks the critical BCS temperature. Blue thin dashed curve (online) marked 50-50, corresponds to perfect symmetry between 2e-/2h-CPs, i.e., the *BCS-Bose crossover*. The blue shaded area (online) is a mixed phase with arbitrary proportions of 2e-/2h-CPs, as well as in the larger Inset below the BCS dot. For $n/n_f < 1$ the pure-phase curves are marked as short-dashed, in this region uncondensed 2h-CPs bosons predominate while in the $n/n_f > 1$ region (pure-phase solid curves) uncondensed 2e-CPs bosons predominate. Symbols (◇, □, ○, △) mark the limit of the two pure-phase GBEC, Crossover and BEC curves, respectively, when $n/n_f \to \infty$, *i.e.*, when $n_f(T) \to 0$, meaning that all unpaired electrons are paired. Smaller Inset shows plot of the dimensionless number density $n_f(T)/n_f$ of unpaired electrons (72) and its $T = 0$ limit n_f, see Ref. [78].

superconductivity. All this rests on four essential ingredients: 1) 2h-CPs cannot and must not be neglected in a fully self-consistent treatment in any many-*fermion* system, otherwise a spurious value of T_c may result that corresponds not to a stable but rather to a *metastable* state; 2) CPs are *bosons*, even though BCS pairs not; 3) CPs are *linearly-dispersive* for small K; 4) to achieve substantially higher T_c's one must depart from the perfect 2e-/2h-CP symmetry of the BCS condensate which in fact is a BE condensate. Neglected in the GBEC theory thus far, however, are: *a*) $K > 0$ terms in the boson-fermion vertex interactions; *b*) boson-boson interactions (as also in BCS theory); *c*) a $T > 0$ Bethe-Salpeter CP treatment; *d*) different hole and electron effective masses; and *e*) ionic-lattice crystallinity effects which might initially be introduced via Van Hove singularities in the electronic DOS or via "bipolarons" instead of CPs.

Lastly, at least two mysteries have surfaced here. In spite of neglecting boson-boson interactions between severely-overlapping CPs, Why has the BCS theory been so successful in describing at least conventional SCs with an ideal gas of Cooper pairs? Why are simple models (such as the BCS or the GBEC theories) quite able to be of any relevance whatsoever in such complex strongly-interacting many-electron systems like SCs?

Acknowledgments

MdeLl thanks V.C. Aguilera-Navarro, A.S. Alexandrov, P.W. Anderson, J.F. Annett, J.R. Clem, J.D. Dow, D.M. Eagles, M. Fortes, M. Grether, B.L. Györffy, T.A. Mamedov, F.J. Sevilla, M.A. Solís, S. Tapia, O. Rojo, J.J. Valencia, A.A. Valladares, H. Vucetich and J.A. Wilson for discussions and/or for providing material prior to its publication. IC thanks CONACyT (Mexico) for postgraduate grant 260978, JO and MdeLl thanks UNAM-DGAPA-PAPIIT (Mexico) for support.

Appendix

A Chemical Potential Limiting Cases

Here we investigate the behaviour of the number density of 2e-CPs and 2h-CPs when fermion chemical potential is bounded by the boson phenomenological energies at zero-CMM.

A1. Limit Case with $2\mu = E_-(0)$

Let us investigate the system of equations $\{(i), (ii), (iii)\}$ which exist only if $E_-(0) \leq E_+(0)$. Let us suppose that we have *strict* inequality $E_-(0) < E_+(0)$. Let us consider at first the limiting case $2\mu = E_-(0)$. In this case the equation (ii) is satisfied but not (i). Thus, $n_0 = 0$ and $m_0 \neq 0$. Putting $n_0 = 0$ and $2\mu = E_-(0)$ in (iii) we get

$$n = -2m_0 + n_-(T) \tag{90}$$

in which, in view of the definitions (35) for $E_{\pm}(K)$, leads to

$$\begin{aligned} n_-(T) &\equiv 2\left(\frac{A_d}{L^d}\right)\int\limits_0^\infty k^{d-1}dk\left(\exp[\beta(\hbar^2k^2/2m\ - E_-(0)/2)]+1\right)^{-1} \\ &+ 2\left(\frac{A_d}{L^d}\right)\int\limits_{0^+}^\infty K^{d-1}dK\left(\exp[\beta(\varepsilon_K + E_+(0) - E_-(0)] - 1\right)^{-1} \\ &- 2\left(\frac{A_d}{L^d}\right)\int\limits_{0^+}^\infty K^{d-1}dK\left(\exp[\beta\varepsilon_K] - 1\right)^{-1}. \end{aligned} \tag{91}$$

By ε_K we imply here the bosonic dispersion (5) cited earlier. Note that the second integral in the right hand side converges because $E_+(0) - E_-(0) > 0$. Thus (90) and (91) must be solved for the density m_0 of 2h CPs, i.e., for the function $m_0 = m_0(T, n)$. As can be seen immediately from (90) and (91) this equation has solutions only in the region of such values of temperature T and the total density n which satisfy the inequality

$$n \le n_-(T) \tag{92}$$

because in (90) we always have $m_0 \ge 0$. This inequality can be written in the alternative form

$$T \le T_-(n) \tag{93}$$

if we designate by T_- the solution of the equation

$$n = n_-(T_-). \tag{94}$$

In other words at fixed value of temperature T the 2h-CP condensate can exist in our BF system only for small enough values of fermion density $n_-(T)$ that are less than some critical density $n_-(0)$. At the given fermion density value $n \le n_-(0)$ this condensate can exist only at temperatures T, which have values between zero and $T_-(n)$. In the region considered, which satisfies the above inequalities, the equations (90) and (91) can be solved trivially and so we get

$$m_0 = \tfrac{1}{2}[n_-(T) - n]. \tag{95}$$

We see that the volume density m_0 of condensed 2h-CPs at fixed temperature T is *linearly decreased* with *increasing* of the fermion density $n_-(T)$ up to the moment when reaches the critical value n (the total number density), at which the density of condensed 2h-CPs becomes zero. At zero temperature $T = 0$ the second and third integrals in (91) become equal to zero, as we have inequality $E_+(0) > E_-(0)$, and the first integral becomes that over the step function $\theta\left[\sqrt{mE_-(0)}/(\hbar - k)\right]$ so that we immediately have

$$n_-(0) = \frac{2A_d}{(L\hbar)^d}\left(\frac{[mE_-(0)]^{d/2}}{d}\right). \tag{96}$$

At low temperatures the first integral in (91) is equal to the quantity (96) approximately with the exactness up to quadratic terms in T. The second integral has exponential smallness because we assume that $E_+(0) > E_-(0)$, which is justified *a posteriori*. Using the linear bosonic dispersion relation, the third integral (91) is $\propto T^d$. This is seen if we replace the integration variable K by $k_B T x / b(d)\lambda\hbar v_F$, after which we obtain for the third integral

$$\int_{0+}^{\infty} K^{d-1} dK \left(\exp[\beta\varepsilon_K] - 1\right)^{-1} = \left(\frac{k_B T}{b(d)\lambda\hbar v_F}\right)^d \int_{0+}^{\infty} \frac{x^{d-1}dx}{\exp[x]-1} = \left(\frac{k_B T}{b(d)\lambda\hbar v_F}\right)^d \Gamma(d)\zeta(d). \tag{97}$$

Here $\Gamma(d)$ is the gamma function, while $\zeta(d)$ is the Riemann Zeta function of order d which for $d = 2$ is $\pi^2/6$. Thus at low temperatures we have the approximate result

$$n_-(T) \simeq \frac{2A_d}{(L\hbar)^d}\left[\frac{[mE_-(0)]^{d/2}}{d} - \frac{(k_B T)^d\, m^{d/2}\,\Gamma(d)\,\zeta(d)}{2^{d/2}\,[b(d)\lambda]^d\, E_F^{d/2}}\right] + \mathcal{O}(T^d) \tag{98}$$

since $E_F = \frac{1}{2}mv_F^2$. For two dimensional systems the first term of eq. (91) can be computed analytically and the first correction to (96) at low temperatures gives

$$\left(\frac{2A_d}{(L\hbar)^2}\right)\left(\frac{m}{kT}\right)\exp\left[-E_-(0)/2kT\right] + \cdots \tag{99}$$

which is always much smaller than T^d for $d = 2$ and has been ignored. On putting (98) in eq. (90), we get the following approximate formula for the threshold temperature

$$k_B T_-(n) \simeq \frac{b(d)\lambda}{[\Gamma(d)\zeta(d)]^{1/d}}\left(\frac{E_F}{m}\right)^{1/2}\left[\frac{[mE_-(0)]^{d/2}}{d} - \frac{\hbar^d n_-(T)}{2(A_d/L^d)}\right]^{1/d} \tag{100}$$

which is accurate at densities n in the vicinity of $n_-(0)$.

A2. Limit Case with $2\mu = E_+(0)$

The equation (i) of system is satisfied, but not (ii). Consequently, $n_0 \neq 0$ and $m_0 = 0$. On putting $m_0 = 0$ and $2\mu = E_+(0)$ in (iii) we get

$$n = 2n_0 + n_+(T) \tag{101}$$

where

$$\begin{aligned} n_+(T) &= 2\left(\frac{A_d}{L^d}\right)\int_0^{+\infty} k^{d-1}dk\left(\exp[\beta\hbar^2k^2/2m - E_+(0)/2] + 1\right)^{-1} \\ &+ 2\left(\frac{A_d}{L^d}\right)\int_{0+}^{\infty} K^{d-1}dK\left(\exp[\beta\varepsilon_K] - 1\right)^{-1} \\ &- 2\left(\frac{A_d}{L^d}\right)\int_{0+}^{\infty} K^{d-1}dK\left(\exp[\beta(\varepsilon_K + E_+(0) - E_-(0))] - 1\right)^{-1}. \end{aligned} \tag{102}$$

Again, the third integral is convergent because $E_+(0) - E_-(0) > 0$. The equation (101),(102) needs solve in respect to the density n_0 of condensed 2e-CPs. From the solution we obtain the function $n_0 = n_0(T, n)$. Obviously this equation has solutions not in the whole region of values of temperature T and of the total density n. For existing of solutions, as it immediately seen from (101), (102), it is necessary that the following inequality

$$n \geq n_+(T) \tag{103}$$

is fulfilled, as we always have $n_0 \geq 0$ in eq.(101). This inequality we can represent also in the following alternative form:

$$T \leq T_+(n) \tag{104}$$

if we designate by T_+ the solution of

$$n = n_+(T_+). \tag{105}$$

At fixed value of temperature T the 2e-CP condensate in our BF system appears only provided that the fermion density $n_+(T)$ exceeds some critical value n. At fixed values of fermion density $n > n_+(0)$ this condensate can exist only at temperatures T lying between zero and a critical temperature $T_+(n)$. In the region considered, which is characterized by these inequalities, the solution of (101), (102) is given by

$$n_0 = \tfrac{1}{2}[n - n_+(T)]. \tag{106}$$

We see that at fixed temperature T the volume density of condensed 2e-CPs is *linearly increasing* with the *increasing* of fermion density $n_+(T)$, beginning from zero up the critical value n. At $T = 0$ the second and third integrals in (102) vanish because $E_+(0) > E_-(0)$, and the first integral is readily evaluated. Thus

$$n_+(0) = 2\left(\frac{A_d}{L^d}\right)\frac{[mE_+(0)]^{d/2}}{d}. \tag{107}$$

At low T the first integral is approximately equal to the quantity (107) with exactness of quadratic terms in T. The third integral in (102) is exponentially small. The second integral in (102) can be calculated with the help of replacement of integration variable: $k_B T x / b(d)\lambda\hbar v_F$. Thus, at low T we have

$$n_+(T) \simeq \frac{2A_d}{(L\hbar)^d}\left[\frac{[mE_+(0)]^{d/2}}{d} - \frac{(k_B T)^d\, m^{d/2}\,\Gamma(d)\zeta(d)}{[b(d)\lambda\hbar]^d\, E_F^{d/2}}\right] + \mathcal{O}(T^d). \tag{108}$$

For two dimensional systems the first term of eq. (102) can be computed analytically and the first correction to (107) at low temperatures gives

$$\left(\frac{2A_d}{(L\hbar)^2}\right)\left(\frac{m}{kT}\right)\exp\left[-E_+(0)/2kT\right] + \cdots \tag{109}$$

which is always much smaller than T^d and has been ignored. On putting (108) in (106) we shall get approximately for the threshold temperature:

$$k_B T_+(n) \simeq \frac{b(d)\lambda}{[\Gamma(d)\zeta(d)]^{1/d}}\left(\frac{E_F}{m}\right)^{1/2}\left[\frac{(mE_+(0))^{d/2}}{d} - \frac{\hbar^d n_+(T)}{2(A_d/L^d)}\right]^{1/d} \tag{110}$$

which is accurate at densities n in the vicinity of the critical density $n_+(0)$.

A3. Limit Case with $E_-(0) < \mu < E_+(0)$

Then neither (i) nor (ii) are satisfied. Thus we need suppose that the third equation (iii) takes the form with $n_0 = 0$ and $m_0 = 0$. Thus the equation (iii) is as follows:

$$\begin{aligned} n &= 2\left(\frac{A_d}{L^d}\right)\int_0^\infty k^{d-1}dk\,(\exp[\beta\xi_k]+1)^{-1} \\ &+ 2\left(\frac{A_d}{L^d}\right)\int_{0^+}^\infty K^{d-1}dK\,(\exp[\beta\{E_+(K)-2\mu\}]-1)^{-1} \\ &- 2\left(\frac{A_d}{L^d}\right)\int_{0^+}^\infty K^{d-1}dK\,(\exp[\beta\{2\mu-E_-(K)\}]-1)^{-1} \end{aligned} \tag{111}$$

where ξ_k, $[E_+(K)-2\mu]$ and $[2\mu - E_-(K)]$ were defined in (35). The equation (111) permits us to find the fermion chemical potential μ as a function of the temperature T and the total number density n. At *zero* temperature the second and the third integrals in (111) vanish because $E_+(0)-\mu > 0$ and $\mu - E_-(0) > 0$. So we have the simple result

$$n = \frac{(m\mu)^{d/2}}{d}\left(\frac{2A_d}{(L\hbar)^d}\right) \tag{112}$$

or

$$\mu = \mu(n) = \frac{\hbar^2}{m}\left[\frac{dn}{2(A_d/L^d)}\right]^{2/d} \tag{113}$$

This is precisely the value of the chemical potential of an ideal fermion gas of density n and at $T = 0$. From (113) we get

$$\mu = E_-(0) \qquad \text{when} \qquad n = n_-(0) = \left(\frac{2A_d}{(L\hbar)^d}\right)\left(\frac{[mE_-(0)]^{d/2}}{d}\right) \tag{114}$$

and

$$\mu = E_+(0) \qquad \text{when} \qquad n = n_+(0) = \left(\frac{2A_d}{(L\hbar)^d}\right)\left(\frac{[mE_+(0)]^{d/2}}{d}\right). \tag{115}$$

All this shows that in the case of complete absence of the BF interaction there occur phase transitions which are the formation and evaporation of 2e- and 2h-CP condensates.

There exist *two* different regions of fermion density $n_\pm(T)$ values. In the region of small fermion densities $n \le n_-(0)$ at enough small temperatures $0 \le T \le T_-(n)$ a pure 2h-CP condensate appears. In the region of large fermion densities $n \ge n_+(0)$ at enough small temperatures $0 \le T \le T_+(n)$ a pure 2e-CP condensate occurs. In the region of intermediate densities: $n_-(0) \le n \le n_+(0)$ there is no condensate at all. Finally, a *mixed condensate* consisting of both 2e- and 2h-CPs *never occurs within the BF system without BF interactions* (i.e., $f \equiv 0$); this contrasts with its existence whenever $f > 0$ as we shall see later. Let us now go from extensive Ω_0' to intensive $\Omega_0'/V \equiv \omega_0'$ quantities. We then obtain from (52) for the modified thermodynamic potential per unit volume

$$\begin{aligned}
\frac{\Omega_0'}{V} &\equiv \omega_0'(T,\mu,n_0,m_0) = [E_+(0)-2\mu]n_0 + [2\mu - E_-(0)]m_0 \\
&+ 2\left(\frac{A_d}{L^d}\right)\int_0^\infty k^{d-1}dk\,(\xi_k - E_k) - 2k_BT\left(\frac{A_d}{L^d}\right)\int_0^\infty k^{d-1}dk\,\ln(1+\exp[-\beta E_k]) \\
&+ 2k_BT\left(\frac{A_d}{L^d}\right)\int_{0+}^\infty K^{d-1}dK\,\ln\left(1-\exp[-\beta(E_+(K)-2\mu)]\right) \\
&+ 2k_BT\left(\frac{A_d}{L^d}\right)\int_{0+}^\infty K^{d-1}dK\,\ln\left(1-\exp[-\beta(2\mu - E_-(K))]\right). \qquad (116)
\end{aligned}$$

The expression for $\Omega_0'/V \equiv \omega_0'$ can be given in a much more convenient form by going from integrations over momenta to integrations over corresponding energies. To do so we define the density of states (DOS) for individual fermion states for one spin per unit volume $N(\epsilon)$ as

$$\left(\frac{A_d}{L^d}\right)\int k^{d-1}dk\cdots \equiv \int d\epsilon N(\epsilon)\cdots \qquad (117)$$

which in turn gives

$$N(\epsilon) \equiv k^{d-1}\left(\frac{A_d}{L^d}\right)\left(\frac{dk}{d\varepsilon}\right) = \frac{2m\epsilon}{\hbar^2}\left(\frac{A_d}{L^d}\right)\left(\frac{m}{\hbar^2}\right)^{d/2-1} \qquad (118)$$

with the last step following since $\epsilon \equiv \hbar^2k^2/2m$. The bosonic DOS per unit volume $M(\varepsilon)$ is similarly defined through (15) as

$$\left(\frac{A_d}{L^d}\right)\int K^{d-1}dK\cdots \equiv \int d\varepsilon M(\varepsilon)\cdots \qquad (119)$$

Thus, using the definitions of A_2 and A_3 given just above we have in 2D

$$M(\varepsilon) \equiv \left(\frac{1}{2\pi}\right)K\left(\frac{dK}{d\varepsilon}\right) \simeq \frac{2\pi\varepsilon}{\lambda^2\hbar^2v_F^2} \qquad (120)$$

where in the last step we used for ε the leading term $b(d)\lambda\hbar v_F K$ in accordance with (5). And in 3D

$$M(\varepsilon) \equiv \left(\frac{1}{2\pi^2}\right)K^2\left(\frac{dK}{d\varepsilon}\right) \simeq \frac{32\varepsilon^2}{\pi^2\lambda^3\hbar^3v_F^3}. \qquad (121)$$

B Condensation Energy at $T \geq 0$

The general expression for the $T \geq 0$ condensation energy per unit volume according to the GBEC theory, is

$$\frac{[F_s(T) - F_n(T)]}{L^d} \equiv \frac{[\Omega_s - \Omega_n]}{L^d} + \frac{[\mu_s - \mu_n]}{L^d} \tag{122}$$

with F the Helmholtz free energy defined by $F \equiv E - TS = \Omega + \mu N$ for any T, and S being the entropy. Since μ is the same for either superconducting s or normal n phases with internal energies E_s and E_n, respectively, in the temperature range $0 \leq T \leq T_c$, one has

$$[F_s(T) - F_n(T)]/L^d = [\Omega_s - \Omega_n]/L^d. \tag{123}$$

To determine the $\Omega(T, L^d, \mu, N_0, M_0)/L^d$ associated with the normal state, i.e., $n_0T) = m_0(T) = 0$ so that $\Delta(T) = 0$, and defining $\epsilon - \mu \equiv \xi$, one obtains [79]

$$\begin{aligned}\Omega_n/L^d &= 2\int_{-\mu}^{0} d\xi N(\xi)\xi - 2k_BT\int_0^\infty d\epsilon N(\epsilon)\ln\left(1+\exp[-\beta|\epsilon-\mu|]\right) \\ &+ 2k_BT\int_{0+}^{\infty} d\varepsilon M(\varepsilon)\ln\left(1-\exp[-\beta(\hbar\omega_D+\varepsilon)]\right).\end{aligned} \tag{124}$$

On the other hand, for BCS theory $n_B(T) = m_B(T)$ implies $n_{B+}(T) = m_{B+}(T)$, it follows from 67) and (53) that one gets $\mu = E_F$. Then, the $\Omega(T, L^d, \mu, N_0, M_0)$ associated with the superconducting state is [79]

$$\begin{aligned}\Omega_s/L^d &= 2n_0\delta\varepsilon + 2\int_{-\mu}^{-\hbar\omega_D} d\xi N(\xi)\xi - 2\int_0^{\hbar\omega_D} d\xi N(\xi)\sqrt{\xi^2+\Delta^2} \\ &- 2k_BT\int_0^\infty d\varepsilon N(\varepsilon)\ln\left(1+\exp[-\beta\sqrt{(\varepsilon-\mu)^2+\Delta^2}\right).\end{aligned} \tag{125}$$

Now, using 124 and 125 and substituting the values in 122 one gets the analytic expression of the condensation energy per unit volume for $T \geq 0$ in the BCS theory, namely [4]

$$\begin{aligned}(\Omega_s - \Omega_n)/L^d &= 2n_0\hbar\omega_D + 2\int_0^{\hbar\omega_D} d\xi N(\xi)\xi - 2\int_0^{\hbar\omega_D} d\xi N(\xi)\sqrt{\xi^2+\Delta^2} \\ &+ 4k_BT\int_0^{\hbar\omega_D} d\xi N(\xi)\ln\left(\frac{1+\exp[-\beta\xi]}{1+\exp[-\beta\sqrt{\xi^2+\Delta^2}]}\right).\end{aligned} \tag{126}$$

According to (122), (126) this can be rewritten in the equivalent form [79]

$$\begin{aligned}[F_s(T) - F_n(T)]/L^d &= (\Omega_s - \Omega_n)/L^d \\ &= \int_0^{\hbar\omega_D} d\xi\Big[2n_0 + N(\xi)\left(\frac{4}{\beta}\right)\ln\left(\frac{1+\exp(-\beta\xi)}{1+\exp\left(-\beta\sqrt{\xi^2+\Delta^2}\right)}\right) \\ &+ 2\left(\xi - \sqrt{\xi^2+\Delta^2}\right)\Big] \equiv \int_0^{\hbar\omega_D} d\xi I(\xi, T).\end{aligned} \tag{127}$$

To show that this is ≤ 0 for all $0 \leq T \leq T_c$ we observe that:

1) $I(\xi, T)$ is a continuous and differentiable function in ξ and is strictly increasing over the integration range (see Appendix A of Ref. [79]). The function $I(\xi, T)$has the following behavior:

   a) $I(\xi = 0, T) = 2n_0 - 2N(\xi)\Delta - 4N(\xi)k_B T \ln\left[(1 + \exp[-\beta\Delta])/2\right] < 0$

   b) $I(\xi = \hbar\omega_D, T) = 2n_0(1 - \lambda) > 0$ and

   c) $I(\xi, T) = 0$ in the cases $\xi = \xi_0 \equiv \lambda\hbar\omega_D$ or $T = T_c$

2) The value of $I(\xi, T)$ vanishes at $T = T_c$ for all $0 \leq \xi \leq \hbar\omega_D$; this implies that the integral (127) is zero. It follows that the Helmholtz free energy for both the superconducting and the normal phase match exactly at this temperature.

3) For $T = 0$ the value of 127 turns out to be

$$\left[F_s(0) - F_n(0)\right]/L^d = N(0)(\hbar\omega_D)^2\left[1 - \sqrt{1 + \Delta_0/\hbar\omega_D)^2}\right] < 0 \tag{128}$$

   where Δ_0 is the gap at $T = 0$. This energy gives a rigorous-upper-bound to the exact ground-state energy of the BCS Hamiltonian.

4) The derivative (see appendix C of Ref. [79])

   a) $\partial I(\xi, T)/\partial T > 0$ for $\xi < \lambda\hbar\omega_D = \xi_0$,

   b) $\partial I(\xi, T)/\partial T < 0$ for $\xi > \lambda\hbar\omega D$,

   c) $\partial I(\xi, T)/\partial T = 0$ for $\xi = \xi_0$.

Following these observations we put

$$\int_{\xi_0}^{\hbar\omega_D} d\xi I(\xi, T) \leq I(\xi = \hbar\omega_D, T)(\hbar\omega_D - \xi_0) = 2n_0(1 - \lambda)(\hbar\omega_D - \xi_0) \equiv R_1 \tag{129}$$

and

$$\int_0^{\xi_0} d\xi I(\xi, T) \leq -|I(\xi', T)|\xi' = 2n_0(\xi' - \lambda\hbar\omega_D) \equiv -R_1 \tag{130}$$

where $\xi' = \left[\lambda - (1 - \lambda)^2\right]\hbar\omega_D \leq \xi_0$, such that $\xi' I(\xi', T) = -R_1$. So, the sum of (129) and (130) gives

$$\int_0^{\hbar\omega_D} d\xi I(\xi, T) \leq 0 \tag{131}$$

Therefore, $\left[F_s(T) - F_n(T)\right]/L^d \leq 0$ in $0 \leq T \leq T_c$. **QED** [79].

C "Electron" and "hole" Fermi and Bose Creation/ Annhilation Operators

C1. Fermions

For single fermions, as for any Fermi-particles, we are able to use two alternative mathematically equivalent *representations* – so-called "electron" and "hole" ones. Thus we can speak instead of "electrons" of "holes," which fermions remain vacant after themselves, as about new individual Fermi-particles.

Indeed, it is always possible to perform the *canonical transformation* (i.e., taking invariant the form of canonical anticommutation relations) of fermion creation and annihilation operators in such a way that we shall consider the creation operator as the annihilation one, and the annihilation operator as the creation one, i.e., we can consider new so-called "hole" Fermi-operators (designate with a special superscript index "h"). Specifically,

$$a^{+h}_{\mathbf{k},s} = a_{\mathbf{k},s} \qquad a^{h}_{\mathbf{k},s} = a^{+}_{\mathbf{k},s}. \tag{132}$$

Obviously, "hole fermion" operators satisfy Fermi anticommutation relations

$$\begin{aligned}
\left\{a^{+h}_{\mathbf{k},s}, a^{+h}_{\mathbf{k}',s'}\right\} &= a^{+h}_{\mathbf{k},s}a^{+h}_{\mathbf{k}',s'} + a^{+h}_{\mathbf{k}',s'}a^{+h}_{\mathbf{k},s} = 0 \\
\left\{a^{h}_{\mathbf{k},s}, a^{h}_{\mathbf{k}',s'}\right\} &= a^{h}_{\mathbf{k},s}a^{h}_{\mathbf{k}',s'} + a^{h}_{\mathbf{k}',s'}a^{h}_{\mathbf{k},s} = 0 \\
\left\{a^{h}_{\mathbf{k},s}, a^{+h}_{\mathbf{k}',s'}\right\} &= a^{h}_{\mathbf{k},s}a^{+h}_{\mathbf{k}',s'} + a^{+h}_{\mathbf{k}',s'}a^{h}_{\mathbf{k},s} = \delta_{\mathbf{k}',\mathbf{k}}\delta_{s',s}.
\end{aligned} \tag{133}$$

The particle and hole occupation number operators are connected with each other by the simple relation

$$\begin{gathered}
\hat{n}^{h}_{\mathbf{k},s} = 1 - \hat{n}_{\mathbf{k},s} \\
\hat{n}^{h}_{\mathbf{k},s} \equiv a^{+h}_{\mathbf{k},s}a^{h}_{\mathbf{k},s} \qquad \hat{n}_{\mathbf{k},s} \equiv a^{+}_{\mathbf{k},s}a_{\mathbf{k},s}.
\end{gathered} \tag{134}$$

Eigenvalues of either operators $\hat{n}_{\mathbf{k},s}$ and $\hat{n}^{h}_{\mathbf{k},s}$ take on values 0 or 1. These eigenvalues are connected with each other by the same relation as operators themselves: because the same one-particle states $\mathbf{k}, s$ are eigenstates for each of them.

If the state $\mathbf{k}, s$ is occupied by an "electron," i.e., $n_{\mathbf{k},s} = 1$, then this state lacks a "hole," i.e., $n^{h}_{\mathbf{k},s} = 0$. And vice versa, if the state $\mathbf{k}, s$ is empty of an "electron," i.e., $n_{\mathbf{k},s} = 0$, then it is occupied by a "hole," i.e., $n^{h}_{\mathbf{k},s} = 1$. This can be easily seen with the application of (134) to the vacuum state and consequently when we change "electron" creation and annihilation operators to "hole" creation and annihilation operators the Hamiltonian of system of noninteracting fermions transforms as

$$\hat{H} \equiv \sum_{\mathbf{k},s} \epsilon_k a^{+}_{\mathbf{k},s}a_{\mathbf{k},s} = -\sum_{\mathbf{k},s} \epsilon_k a^{+h}_{\mathbf{k},s}a^{h}_{\mathbf{k},s} \tag{135}$$

and the total number of fermions as

$$\hat{N} \equiv \sum_{\mathbf{k},s} a^{+}_{\mathbf{k},s}a_{\mathbf{k},s} = -\sum_{\mathbf{k},s} a^{+h}_{\mathbf{k},s}a^{h}_{\mathbf{k},s}. \tag{136}$$

Note the appearance of minus signs on both right hand sides.

Thus, fermions as well as any Fermi-particles in the second quantization representation can be described equally well in "electron" or in the "hole" representation; both descriptions are mathematically equivalent. In the present paper, following tradition, we will use only "electron" representation for fermions, not "hole" one.

C2. Bosons

For 2e or 2h , as for any Bose particles, as well as for Fermi particles, we can also use one of two mathematically equivalent alternative representations, "electron" or "hole."

In contrast with Fermi-particles, however, now "hole Bose operators" do *not* satisfy ordinary Bose commutation relations, and for this reason we must talk now about "hole bosons" as a special kind of individual quantum particles. The transformation from "electron" to "hole" bosons is *not* canonical now (i.e., does *not* conserve invariant the commutation relations for creation and annihilation operators).

Let us suppose that he creation "electron" operator is precisely the annihilation "hole" operator, and vice versa, namely

$$c^{\dagger}_{\mathbf{K}} = b_{\mathbf{K}} \qquad c_{\mathbf{K}} = b^{\dagger}_{\mathbf{K}} \tag{137}$$

where we have used the letter "c" to designate "hole Bose operators". These operators satisfy the commutation relations

$$\begin{aligned} \left[c^{\dagger}_{\mathbf{K}}, c^{\dagger}_{\mathbf{K}'}\right] &= c^{\dagger}_{\mathbf{K}}c^{\dagger}_{\mathbf{K}'} - c^{\dagger}_{\mathbf{K}'}c^{\dagger}_{\mathbf{K}} = 0 \\ \left[c_{\mathbf{K}}, c_{\mathbf{K}'}\right] &= c_{\mathbf{K}}c_{\mathbf{K}'} - c_{\mathbf{K}'}c_{\mathbf{K}} = 0 \\ \left[c_{\mathbf{K}}, c^{\dagger}_{\mathbf{K}'}\right] &= c_{\mathbf{K}}c^{\dagger}_{\mathbf{K}'} - c^{\dagger}_{\mathbf{K}'}c_{\mathbf{K}} = -\delta_{\mathbf{K}',\mathbf{K}} \end{aligned} \tag{138}$$

which we obtain immediately from the ordinary Bose commutation relations valid for operators $b^{+}_{\mathbf{K}}$ and $b_{\mathbf{K}}$. In contrast with Bose relations the commutation relations now obtained (for "hole Bose-operators") have *minus signs* before Kronecker symbol $\delta_{\mathbf{K}',\mathbf{K}}$ on the right hand side of the last equation. This means in particular that occupation numbers of hole Bose operators $\hat{M}_{\mathbf{K}} = c^{\dagger}_{\mathbf{K}}c_{\mathbf{K}}$ have negative eigenvalues: $0, -1, -2, \cdots - \infty$.

For the Hamiltonian of noninteracting hole bosons and for the operator of total number of them we have formulas:

$$\begin{aligned} \hat{H} &= \sum_{\mathbf{K}} E_{-}(K)c^{\dagger}_{\mathbf{K}}c_{\mathbf{K}} \\ \hat{M} &= \sum_{\mathbf{K}} c^{\dagger}_{\mathbf{K}}c_{\mathbf{K}} \end{aligned} \tag{139}$$

where $E_{-}(K) = E_{-}(0) - \hbar^2K^2/2(2m)$, with $\hat{M}$ the hole boson number particle operator and $E(0)$ is the energy of *motionless* or center-momentum-of-mass of hole boson (positive or negative), m is the effective mass of hole Boson which itself is *negative* (because hole Boson has *negative kinetic energy*).

Eigenvalues of operators $\hat{H}$ and $\hat{N}$ we can write as follows:

$$E_{\ldots N_{\mathbf{K}}\ldots} = \sum_{\mathbf{K}} E_{+}(K)N_{\mathbf{K}}, \qquad N_{\ldots N_{\mathbf{K}}\ldots} = \sum_{\mathbf{K}} N_{\mathbf{K}}. \tag{140}$$

where $N_{\mathbf{K}}$ is the electron boson number particle operator, this implies that $M_{\mathbf{K}} = -N_{\mathbf{K}}$, these formulas (140) become

$$E_{...M_{\mathbf{K}}...} = -\sum_{\mathbf{K}} E_-(K) M_{\mathbf{K}}, \qquad M_{...M_{\mathbf{K}}...} = -\sum_{\mathbf{K}} M_{\mathbf{K}}. \tag{141}$$

Thus in "electron" representation of "hole bosons"' for their Hamiltonian and the operator of total number of particles we have

$$\hat{H} = \sum_{\mathbf{K}} \varepsilon_K c^{\dagger}_{\mathbf{K}} c_{\mathbf{K}} \qquad \hat{M} = -\sum_{\mathbf{K}} c^{\dagger}_{\mathbf{K}} c_{\mathbf{K}} \tag{142}$$

where $\varepsilon_K = -E_-(K) = E(0) + \hbar^2 K^2/2(2m)$. "Electron" representation of Bose operators in this paper we use both for $2h$ and for 2e.

References

[1] Ogg, R.A. Jr. Phys Rev 69 (1946) 243.

[2] Ginzburg, V.L. Usp Fiz Nauk 48 (1952) 25; Fortschr Phys **1** (1953) 101.

[3] Schafroth, M.R. Phys Rev 96 (1954) 1442.

[4] Schafroth, M.R. Phys Rev 100 (1955) 463.

[5] Blatt, J.M.; Butler S.T. Phys Rev 100 (1955) 476.

[6] Schafroth, M.R.; Butler, S.T.; Blatt J.M. Helv Phys Acta 30 (1957) 93.

[7] Schafroth, M.R. Sol State Phys 10 (1960) 293.

[8] Blatt, J.M.; Matsubara, T. Prog Theor Phys 20 (1958) 533; 23 (1960) 447 and 451; 27 (1962) 1137.

[9] Blatt, J.M. Theory of superconductivity (Academic Press, New York, 1964).

[10] Bardeen, J.; Cooper, L.N.; Schrieffer, J.R. Phys Rev 108 (1957) 1175.

[11] Bogoliubov, N.N. JETP 34 (1958) 41.

[12] Valatin, J.G. Nuovo Cimmento 7 (1958) 843.

[13] Bogoliubov, N.N.; Tolmachev, V.V.; Shirkov, D.V. Fortschr Phys 6 (1958) 605; and also in A new method in the theory of superconductivity (Consultants Bureau, NY, 1959).

[14] Bardeen, J. Phys Today 16 (1963) 19.

[15] Cooper, L.N. Am J Phys 28 (1960) 91.

[16] Rickayzen, G. Theory of superconductivity (Interscience, New York, 1965).

[17] Ketterson, J.B.; Song, S.N. Superconductivity (Cambridge Universtity Press, Cambridge, 1999).

[18] Cooper, L.N. Phys Rev 104 (1956) 1189.

[19] Ranninger, J.; Robaszkiewicz, S. Physica B 135 (1985) 468.

[20] Ranninger, J.; Micnas, R.; Robaszkiewicz, S. Ann Phys Fr 13 (1988) 455.

[21] Friedberg, R.; Lee, T.D. Phys Rev B 40 (1989) 6745.

[22] Friedberg, R.; Lee, T.D.; Ren, H.C. Phys Rev B 42 (1990) 4122.

[23] Friedberg, R.; Lee, T.D.; Ren, H.C. Phys Lett A 152 (1991) 417 and 423.

[24] Friedberg, R.; Lee, T.D.; Ren, H.C. Phys Rev B 45 (1992) 10732.

[25] Ren, H.C. Physica C 303 (1998) 115.

[26] Tolmachev, V.V. Phys Lett A 266 (2000) 400.

[27] Llano, M. de; Tolmachev, V.V. Physica A 317 (2003) 546.

[28] Leggett, A.J. Nature Physics 2 (2006) 134.

[29] Ehrenfest, P.; Oppenheimer, J.R. Phys Rev 37 (1931) 333.

[30] Nozières, P.; Schmitt-Rink, S. J Low Temp Phys 59 (1985) 195.

[31] Batle, J.; Casas, M.; Fortes, M.; Llano, M. de; Sevilla, F.J.; Solís, M.A.; Tolmachev, V.V. Cond Mat Th 18 (2003) 111.

[32] Casas, M.; Rigo, A.; Llano, M. de; Rojo, O.; Solís, M.A.; Phys. Lett. A 245 (1998) 5.

[33] Casas, M.; Davidson, N.J.; Llano, M. de; Mamedov, T.A.; Puente, A.; Quick, R.M.; Rigo, A.; Solís, M.A. Physica A 295 (2001) 146.

[34] Casas, M.; Llano, M. de; Puente, A.; Rigo, A.; Solís, M.A.; Sol State Comm 123 (2002) 101.

[35] Deaver, B.S. Jr.; Fairbank, W.M. Phys Rev Lett 7 (1961) 43.

[36] Doll, R.; Näbauer, M. Phys Rev Lett 7 (1961) 51.

[37] Gough, C.E.; Colclough, M.S.; Forgan, E.M.; Jordan, R.G.; Keene, M.; Muirhead, C.M.; Rae, I.M.; Thomas, N.; Abell, J.S.; Sutton, S. Nature 326 (1987) 855.

[38] Schneider, C.W.; Hammerl, G.; Logvenov, G.; Kopp, T,; Kirtly, J.R.; Hirschfeld, P.J.; Mannhart, J. Europhys Lett 68 (2004) 86.

[39] Aligia, A.A.; Kampf, A.P.; Mannhart, J. Phys Rev Lett 94 (2005) 247004.

[40] Koh, S.I. Physica C 191 (1992) 167.

[41] Koh, S.I. Phys Rev 49 (1994) 8983.

[42] Tarasewicz, P. Eur Phys J B 41 (2004) 185.

[43] Bunkov, Y.M.; Chen, A.S.; Cousins, D.J.; Godfrin, H. Phys Rev Lett 85 (2000) 3456.

[44] Gough, C. private communications.

[45] Bednorz, J.G.; Müller, K.A. Z Phys 64 (1986) 189.

[46] Gao, L.; Xue, Y.Y.; Chen, F.; Xiong, Q.; Meng, R.L.; Ramírez, D.; Chu, C.W.; Eggert, J.H.; Mao, H.K. Phys Rev B 50 (1994) 4260.

[47] Papaconstantopoulos, D.A.; Klein, B.M.; Mehl, M.J.; Pickett, W.E. Phys Rev B 91 (2015) 184511.

[48] Fortes, M.; Llano, M. de; Solís, M.A. to be published.

[49] Aguilera-Navarro, V.C.; Fortes, M.; Llano, M. de Sol St Comm 129 (2004) 577.

[50] Fortes, M.; Solís, M.A.; Llano, M. de; Tolmachev, V.V. Physica C 364 (2001) 95.

[51] Hirsch, J.E. Physica C 341-348 (2000) 213 and refs. therein.

[52] Verheijen, A.A.; van Ruitenbeek, J.M.; de Bruyn Ouboter, R.; de Jongh, L.J.; Nature 345 (1990) 418.

[53] Hirsch, J.E. Phys Rev B 68 (2003) 012510.

[54] Colino, J.; Gonzalez, M.A.; Martín, J.I.; Velez, M.; Oyola, D.; Prieto, P.; Vicent, J.L. Phys Rev B 49 (1994) 3496.

[55] Jin, R.; Ott, H.R. Phys Rev B 57 (1998) 13872.

[56] Göb, W.; Liebich, W.; Lang, W.; Puica, I.; Sobolewski, R.; Rössler, R.; Pedarnig, J.D.; Büerle, D. Phys Rev B 62 (2000) 9780 and refs. therein.

[57] Bogoliubov, N.N.; Nuovo Cimmento 7 (1958) 794.

[58] Mamedov, T.A.: Llano, M. de J Phys Soc Japan 79 (2010) 044706; 80 (2011) 074718; Phil Mag 93 (2013) 2896 and 94 (2014) 4102.

[59] Valatin, J.; Nuovo Cimmento 7 (1958) 843.

[60] Corson, J.; Mallozo, R.; Orenstein, J.; Eckstein, J.N.; Bozovic, I. Nature 398 (1999) 221.

[61] Corson, J.; Orenstein, J.; Oh, S.; O'Donnell, J.; Eckstein, J.N. Phys Rev Lett 85 (2000) 2569.

[62] Labbé, J.; Barisic, F.; Friedel, J. Phys Rev Lett 19 (1967) 1039.

[63] Eagles, D.M. Phys Rev 186 (1969) 456.

[64] Leggett, A.J. J Phys (Paris) Colloq 41 (1980) C7-19.

[65] Miyake, K. Prog Theor Phys 69 (1983) 1794.

[66] Nozières, P.; Schmitt-Rink, S. J Low Temp Phys 59 (1985) 195.

[67] Randeria, M.; Duan, J.M.; Shieh, L.Y.; Phys Rev Lett 62 (1989) 981.

[68] van der Marel, D. Physica C 165 (1990) 35.

[69] Bar-Yam, Y. Phys Rev B 43 (1991) 359 and 2601.

[70] Drechsler, M.; Zwerger, W. Ann der Physik 1 (1992) 15.

[71] Haussmann, R. Z Phys B 91 (1993) 291.

[72] Pistolesi, F.; Strinati, G.C. Phys Rev B 53 (1996) 15168.

[73] Adhikari, S.K.; Llano, M. de; Sevilla, F.J.; Solís, M.A.; Valencia, J.J. Physica C 453 (2007) 37.

[74] Kittel, C. Introduction to solid state physics (John Wiley & Sons, USA, 1996) p. 343.

[75] Annett, J.F. Superconductivity, superfluids and condensates (Oxford University Press, New York, 2004).

[76] Pahtria, R.K. Statistical mechanics (Butterworth-Heinemann, Oxford, 1996) p. 176.

[77] Uemura, Y.J. J Phys Cond Mat 16 (2004) S4515 and more recently in Physica B, 1 (2006) 374.

[78] Chávez, I.; Grether, M.; Llano, M. de J Supercond Novel Magn 28 (2015) 309.

[79] Ortega, J.; Zúñiga, F.; Llano, M. de to be published.

In: Superconductivity
Editor: Paulette Grant

ISBN: 978-1-63483-907-5

Chapter 5

KINEMATICAL SPIN-FLUCTUATION MECHANISM OF HIGH-TEMPERATURE SUPERCONDUCTIVITY IN CUPRATES

***Nikolay M. Plakida*[1,*] *and Viktor S. Oudovenko*[2]**
[1]Joint Institute for Nuclear Research, Dubna, Russia
[2]Rutgers University, New Jersey, US

Abstract

We propose a microscopic theory of superconductivity based on the spin-fluctuation mechanism of pairing. We argue that the cuprate superconductors are strongly correlated electronic systems where electronic excitations have to be described by the projected (Hubbard) electron operators in the singly- and doubly occupied Hubbard subbands. Due to the unconventional commutation relations for the Hubbard operators, a specific kinematical interaction of electrons with spin and charge fluctuations emerges. A new energy scale of the order of kinetic energy of electrons W appears in the intraband hopping which is much larger than the antiferromagnetic exchange interaction J induced by the interband hopping. Taking into account this interaction, a microscopic theory of superconductivity in cuprates is presented within the extended Hubbard model where the intersite Coulomb repulsion V and electron-phonon interaction are taken into account. The Dyson equation for the normal and pair Green functions for two Hubbard subbands is derived using Mori-type projection technique. The self-energy is calculated in the self-consistent Born approximation. We found the d-wave pairing with high-T_c mediated by the kinematical spin-fluctuation interaction. Contributions coming from the realistic in cuprates intersite Coulomb repulsion and electron-phonon interaction turned out to be small since only the angular momentum $l = 2$ of interactions gives contributions to the d-wave pairing. Isotope effect on T_c induced by electron-phonon interaction is weak at optimal doping and increased at low doping. Superconductivity can be suppressed only for a large intersite Coulomb repulsion $V > W$. The kinematical interaction is absent in the spin-fermion models and is lost in the slave-boson (-fermion) models treated in the mean-field approximation.

*E-mail address: plakida@theor.jinr.ru

PACS: 74.20.Mn, 71.27.+a, 71.10.Fd, 74.72.-h

Keywords: nonconventional mechanisms, strongly correlated electron systems, heavy fermions, lattice fermion models (Hubbard model, etc.), cuprate superconductors, high-temperature superconductivity, strongly correlated electronic systems, Hubbard model, cuprates

1. Introduction

Discovery of high-temperature superconductivity (HTSC) in cuprates by Bednorz and Müller [1] caused an unprecedented wave of scientific activity in the study of this extraordinary phenomenon (see, e.g. [2, 3]). In recent years investigations of high-quality samples using new experimental technique have afforded to describe unconventional physical properties of cuprates with high precision. At the same time, intensive studies of various microscopical models by modern theoretical methods have not so far resulted in a commonly accepted interpretation of all the physical phenomena and the mechanism of superconducting pairing in cuprates. The most frequently discussed scenarios are the conventional electron-phonon pairing (see, e.g., [4, 5]) and the spin-fluctuation mechanism [6] in the framework of phenomenological spin-fermion models (see, e.g., [7, 8, 9, 10] and references therein). The main argument against the spin-fluctuation pairing, the weak intensity of spin fluctuations at the optimal doping seen in magnetic inelastic neutron scattering experiments [11], was rejected in the recent resonant inelastic x-ray scattering experiments [12, 13]. In a large family of cuprate superconductors, paramagnon AF excitations with the dispersion and spectral weight similar to those of magnons in undoped cuprates were observed. Using the magnetic spectrum found in the $YBCO_7$ crystal, superconductivity with $T_c = 100 - 200$ K was predicted [12]. Thus, spin fluctuations have sufficient strength to mediate HTSC in cuprates and to explain various physical properties of cuprate materials as, e.g., the "kink" phenomenon observed in the electronic spectrum by the angle-resolved photoemission spectroscopy [14], the optical and dc conductivity [15], etc. Therefore, it can be suggested that the electron-phonon pairing mechanism plays a secondary role in the cuprate superconductors which is supported by observation of the weak isotope effect in optimally doped samples (see, e.g., reviews [16, 17, 18] and references therein).

In early studies of the problem, a model of strongly correlated electrons was proposed by Anderson [19] where superconductivity occurs at finite doping due to the antiferromagnetic (AF) superexchange interaction. Following this idea, various theoretical models of pairing "from repulsion" were considered such as the conventional Hubbard model [20]. In these studies the intersite Coulomb interaction (CI) V beyond the Hubbard single-site CI U were usually neglected, as e.g., in the renormalization group calculations in Ref. [21], though it is well known that the CI is detrimental to superconductivity. In particular, in Ref. [22] it was stressed that a contribution from the repulsive well-screened weak CI in the first order strongly suppresses the pairing induced by contributions of higher orders. To overcome this critical statement, the extended Hubbard model with intersite CI V was studied in Ref. [23]. The superconducting pairing of various symmetries, extended s-, p-, and d-wave types was found depending on the electron concentration and V. Following the original idea of Kohn-Luttinger [24], in Refs. [25, 26] it was shown that the p-wave superconductivity exists in the electronic gas at low density with a strong repulsion U and a relatively strong intersite CI V. Subsequent studies of the extended Hubbard model in the weak correlation limit have shown that superconducting pairing of different types of symmetry can occur depending on the CI between the nearest V_1 and next V_2 neighbor sites and electron hopping parameters between distant sites in a broad region of electron concentration [27].

In these investigations the Fermi-liquid model in the weak correlation limit was used. We argue that the cuprate superconductors are the Mott-Hubbard (more accurately, charge-transfer) doped insulators where electronic excitations have to be described by the projected (Hubbard) electron operators in the singly- and doubly occupied Hubbard subbands for CI U larger than the electronic kinetic energy W (for reviews see [28, 29]). To elucidate the problem of "pairing from repulsion" in cuprates analytical studies of strongly correlated electronic systems are desirable. A rigorous analytical method is based on the Hubbard operator (HO) technique where the HO algebra is implemented rigorously [30]. Due to the unconventional commutation relations for the HOs, a specific *kinematical interaction* of electrons with spin and charge fluctuations emerges. The superconducting pairing induced by the kinematical interaction was first proposed by Zaitsev and Ivanov [31] who studied the two-particle vertex equation by applying the diagram technique for HOs. The momentum-independent s-wave pairing was found in the lowest order diagram approximation equivalent to the mean-field approximation (MFA). However, this solution violates the HO kinematics as discussed below and the AF exchange interaction or higher order diagrams should be considered to obtain the d-wave pairing (see, e.g., Refs. [32, 33, 34, 35]).

In a series of papers we have developed a microscopic theory of electronic spectrum and superconductivity in the Hubbard model using HO representation (see Refs. [36, 37, 38, 39, 40]). In recent studies [39, 40]) we consider the extended Hubbard model with the intersite CI V in the limit of strong correlations, $U > W$. The kinematical interaction for HOs brings about the weak AF exchange interaction J due to interband hopping and at the same time intraband hopping results in a much stronger kinematical interaction $g_{sf} \sim W \gg J$ of electrons with spin and charge excitations. Therefore, the exchange interaction J that determines spin excitation spectrum is not so important for the pairing interaction driven by g_{sf}. Moreover, we show that the intersite CI V, being in cuprates of the order of J, destroys superconducting pairing in the resonating valence bond (RVB) state of Anderson [19]. To compare various contributions to the superconducting d-wave pairing, we consider also a model of electron-phonon interaction (EPI) with strong forward scattering proposed in Ref. [41]. Contributions coming from weak CI V and phonons turned out to be small since only the angular momentum $l = 2$ of the interactions gives a contribution to the d-wave pairing. We calculate the doping dependence of superconducting T_c for various values of U and V and show that as long as V does not exceed the kinematical interaction, $V \lesssim W$, the d-wave pairing is preserved.

In calculations we use the Mori-type projection technique [42, 43] in the equation of motion method for thermodynamic Green functions (GFs) [44] expressed in terms of the HOs. The Dyson equation for the normal and anomalous (pair) GFs is derived where the self-energy is calculated in the self-consistent Born approximation (SCBA). We show that the kinematical interaction is responsible for strong electron coupling with spin-fluctuations and results in the d-wave superconductivity with high-T_c. The isotope effect on T_c caused by EPI in the extended Hubbard model appears to be weak at optimal doping and increases at low doping as observed in experiments.

In the next section we discuss the kinematical interaction in the extended Hubbard model. In Sec. 3 the Dyson equation is derived and a calculation of the self-energy is given. A self-consistent system of equations is formulated in Sec. 4 where results of computations of the electronic spectrum in the normal state, of superconducting T_c and the d-wave gap function are presented. The isotope effect is discussed in Sec. 4.4. Concluding remarks are given in Sec. 5 Details of the calculations are provided in the Appendix.

2. Kinematical Interaction in the Hubbard Model

To describe electronic systems with strong correlations the Hubbard model is commonly used [20]. We consider the extended Hubbard model on a square lattice

$$H = \sum_{i \neq j,\sigma} t_{ij}\, a^{\dagger}_{i\sigma} a_{j\sigma} - \mu \sum_i N_i + \frac{U}{2} \sum_i N_{i\sigma} N_{i\bar{\sigma}} + H_{c,ep}, \tag{1}$$

where $t_{i,j}$ is the single-electron hopping parameters, $a^{\dagger}_{i\sigma}$ and $a_{i\sigma}$ are the Fermi creation and annihilation operators for electrons with spin $\sigma/2$ ($\sigma = \pm 1 = (\uparrow, \downarrow)$, $\bar{\sigma} = -\sigma$) on the lattice site i, U is the on-site CI. $N_i = \sum_\sigma N_{i\sigma}$, $N_{i\sigma} = a^{\dagger}_{i\sigma} a_{i\sigma}$ is the number operator and μ is the chemical potential. The inertsite CI V_{ij} and EPI g_{ij} are defined by the Hamiltonian

$$H_{c,ep} = \frac{1}{2} \sum_{i \neq j} V_{ij} N_i N_j + \sum_{i,j} g_{ij} N_i\, u_j, \tag{2}$$

where u_j describes an atomic displacement on the lattice site j for a particular phonon mode. More generally, the EPI can be written as a sum $\sum_\nu g^{\nu}_{i,j} u^{\nu}_j$ over the normal modes ν. Note, that much earlier Bogolubov [45] has formulated a general polar model of metals for electrons with Coulomb correlations which is reduced to the Hubbard model if only the largest on-site CI U is taken into account.

In the strong correlation limit, $U \gg |t_{ij}|$, the projected electron operators referring to the singly and doubly occupied subbands, the HOs should be introduced [30]: $a^{\dagger}_{i\sigma} = a^{\dagger}_{i\sigma}(1 - N_{i\bar{\sigma}}) + a^{\dagger}_{i\sigma} N_{i\bar{\sigma}} \equiv X^{\sigma 0}_i + X^{2\bar{\sigma}}_i$. In terms of the HOs the model (1) reads

$$\begin{aligned} H &= \varepsilon_1 \sum_{i,\sigma} X^{\sigma\sigma}_i + \varepsilon_2 \sum_i X^{22}_i + \sum_{i \neq j,\sigma} t_{ij}\, \{X^{\sigma 0}_i X^{0\sigma}_j \\ &+ X^{2\sigma}_i X^{\sigma 2}_j + \sigma\,(X^{2\bar{\sigma}}_i X^{0\sigma}_j + \text{H.c.})\} + H_{c,ep}, \end{aligned} \tag{3}$$

where $\varepsilon_1 = -\mu$ is the single-particle energy, $\varepsilon_2 = U - 2\,\mu$ is the two-particle energy. The number operator and the spin operators in terms of HOs are defined as

$$N_i = \sum_\sigma X^{\sigma\sigma}_i + 2X^{22}_i, \tag{4}$$

$$S^{\sigma}_i = X^{\sigma\bar{\sigma}}_i, \quad S^z_i = (\sigma/2)\,(X^{\sigma\sigma}_i - X^{\bar{\sigma}\bar{\sigma}}_i). \tag{5}$$

In the present paper the model (3) will be used to consider the hole-doped case where the chemical potential μ depends on the average hole occupation number

$$n = 1 + \delta = \langle\, N_i \rangle, \tag{6}$$

where $\langle ... \rangle$ denotes the statistical average with the Hamiltonian (3).

The HO $X_i^{\alpha\beta} = |i\alpha\rangle\langle i\beta|$ describes transition from the state $|i,\beta\rangle$ to the state $|i,\alpha\rangle$ on the lattice site i where (α,β) refer to four possible states: an empty state $(\alpha,\beta = 0)$, a singly occupied state $(\alpha,\beta = \sigma)$, and a doubly occupied state $(\alpha,\beta = 2)$. From the multiplication rule for the HOs $X_i^{\alpha\beta} X_i^{\gamma\delta} = \delta_{\beta\gamma} X_i^{\alpha\delta}$ follows their commutation relations

$$\left[X_i^{\alpha\beta}, X_j^{\gamma\delta}\right]_{\pm} = \delta_{ij}\left(\delta_{\beta\gamma} X_i^{\alpha\delta} \pm \delta_{\delta\alpha} X_i^{\gamma\beta}\right), \tag{7}$$

with the upper sign for the Fermi-type operators (such as $X_i^{0\sigma}$) and the lower sign for the Bose-type operators (such as the number or spin operators). The HOs obey the completeness relation

$$X_i^{00} + \sum_{\sigma} X_i^{\sigma\sigma} + X_i^{22} = 1, \tag{8}$$

which rigorously preserves the constraint that only one quantum state α can be occupied on any lattice site i.

The unconventional commutation relations (7) for HOs result in the kinematical interaction. The term was introduced by Dyson in a general theory of spin-wave interactions [46]. It was pointed out that the spin-wave creation and annihilation operators $b_i^{\dagger} = S_i^{-}$ and $b_i = S_i^{+}$ obey the commutation relation

$$b_i b_j^{\dagger} - b_j^{\dagger} b_i = \delta_{i,j}\,(1 - 2\, b_i^{\dagger} b_i), \tag{9}$$

and therefore, they are Bose operators on different lattice site and Fermi operators on the same lattice site. The last term $n_i = b_i^{\dagger} b_i$ in Eq. (9) just brings about the kinematical interaction which may be small for low density of spin excitations, $n_i \ll 1$. Similar equation can be written for the electron creation $X_i^{\sigma 0}$ and annihilation $X_j^{0\sigma}$ operators,

$$X_i^{0\sigma} X_j^{\sigma 0} + X_j^{\sigma 0} X_i^{0\sigma} = \delta_{i,j}(1 - X_i^{\bar\sigma\bar\sigma} - X_i^{22}), \tag{10}$$

where the relation (8) was used. Therefore, these electron operators with spin σ in the singly occupied subband can be considered as Fermi operators if the number of electrons $X_i^{\bar\sigma\bar\sigma}$ with opposite spins $\bar\sigma$ and doubly occupied states X_i^{22} are small. For finite occupation of these states the last terms in Eq. (10) lead to the kinematical interaction.

To demonstrate the role of the kinematical interaction let us consider the equation of motion for the HO $X_i^{\sigma 2} = a_{i\sigma}^{\dagger} a_{i\sigma} a_{i\bar\sigma}$ in the Heisenberg representation ($\hbar = 1$) for the model (3):

$$\begin{aligned} i\frac{d}{dt} X_i^{\sigma 2} &= [X_i^{\sigma 2}, H] = (\varepsilon_1 + U)\, X_i^{\sigma 2} + \sum_{l,\sigma'} t_{il}\left(B_{i\sigma\sigma'}^{22} X_l^{\sigma' 2} - \sigma B_{i\sigma\sigma'}^{21} X_l^{0\bar\sigma'}\right) \\ &- \sum_{l} t_{il} X_i^{02}\left(X_l^{\sigma 0} + \sigma X_l^{2\bar\sigma}\right) + \sum_{l} X_i^{\sigma 2}(V_{il}\, N_l + g_{il}\, u_l). \end{aligned} \tag{11}$$

Here $B^{\alpha\beta}_{i\sigma\sigma'}$ are the Bose-type operators,

$$\begin{aligned} B^{22}_{i\sigma\sigma'} &= (X_i^{22} + X_i^{\sigma\sigma})\,\delta_{\sigma'\sigma} + X_i^{\sigma\bar\sigma}\,\delta_{\sigma'\bar\sigma} \qquad (12)\\ &= (N_i/2 + \sigma\,S_i^z)\,\delta_{\sigma'\sigma} + S_i^{\sigma}\,\delta_{\sigma'\bar\sigma}, \\ B^{21}_{i\sigma\sigma'} &= (N_i/2 + \sigma S_i^z)\,\delta_{\sigma'\sigma} - S_i^{\sigma}\,\delta_{\sigma'\bar\sigma}, \qquad (13)\end{aligned}$$

where we used the definition of the number operator (4) and the spin operators (5). The last term in (11) is caused by the dynamic intersite CI and the EPI (2). We see that the hopping amplitudes here depend on the number and spin operators due to the kinematical interaction of electrons with spin and charge fluctuations. In phenomenological spin-fermion models, a dynamical coupling of electrons with spin fluctuations is specified by fitting parameters, while in Eq. (11) the interaction is determined by the hopping energy t_{ij} fixed by the electronic dispersion.

3. General Formulation

3.1. Dyson Equation

To consider superconducting pairing in the model (3), we introduce the two-time thermodynamic GF [44] expressed in terms of the four-component Nambu operators, $\hat{X}_{i\sigma}$ and $\hat{X}^{\dagger}_{i\sigma} = (X_i^{2\sigma}\; X_i^{\bar\sigma 0}\; X_i^{\bar\sigma 2}\; X_i^{0\sigma})$:

$$\mathsf{G}_{ij\sigma}(t-t') = -i\theta(t-t')\langle\{\hat{X}_{i\sigma}(t), \hat{X}^{\dagger}_{j\sigma}(t')\}\rangle \equiv \langle\langle \hat{X}_{i\sigma}(t) \mid \hat{X}^{\dagger}_{j\sigma}(t')\rangle\rangle, \qquad (14)$$

where $\{A, B\} = AB + BA$, $A(t) = \exp(iHt)A\exp(-iHt)$, and $\theta(x) = 1$ for $x > 0$ and $\theta(x) = 0$ for $x < 0$. The Fourier representation in $(\mathbf{k}, \omega)$-space is defined by the relations:

$$\mathsf{G}_{ij\sigma}(t-t') = \frac{1}{2\pi}\int_{-\infty}^{\infty} dt e^{-i(t-t')}\mathsf{G}_{ij\sigma}(\omega), \qquad (15)$$

$$\mathsf{G}_{ij\sigma}(\omega) = \frac{1}{N}\sum_{\mathbf{k}} \exp[i\mathbf{k}(\mathbf{i}-\mathbf{j})]\mathsf{G}_{\sigma}(\mathbf{k}, \omega). \qquad (16)$$

The GF (16) is convenient to write in the matrix form

$$\mathsf{G}_{\sigma}(\mathbf{k}, \omega) = \begin{pmatrix} \hat{G}_{\sigma}(\mathbf{k}, \omega) & \hat{F}_{\sigma}(\mathbf{k}, \omega) \\ \hat{F}^{\dagger}_{\sigma}(\mathbf{k}, \omega) & -\hat{G}_{\bar\sigma}(-\mathbf{k}, -\omega) \end{pmatrix}, \qquad (17)$$

where the normal $\hat{G}_{\sigma}(\mathbf{k}, \omega)$ and anomalous (pair) $\hat{F}_{\sigma}(\mathbf{k}, \omega)$ GFs are 2×2 matrices for two Hubbard subbands:

$$\hat{G}_{\sigma}(\mathbf{k}, \omega) = \langle\langle \begin{pmatrix} X_{\mathbf{k}}^{\sigma 2} \\ X_{\mathbf{k}}^{0\bar\sigma} \end{pmatrix} \mid X_{\mathbf{k}}^{2\sigma} X_{\mathbf{k}}^{\bar\sigma 0}\rangle\rangle_{\omega}, \qquad (18)$$

$$\hat{F}_{\sigma}(\mathbf{k}, \omega) = \langle\langle \begin{pmatrix} X_{\mathbf{k}}^{\sigma 2} \\ X_{\mathbf{k}}^{0\bar\sigma} \end{pmatrix} \mid X_{-\mathbf{k}}^{\bar\sigma 2} X_{-\mathbf{k}}^{0\sigma}\rangle\rangle_{\omega}. \qquad (19)$$

To calculate the GF (14) we use the equation of motion method. Differentiating the GF with respect to time t we obtain the equation for the Fourier representation (15)

$$\omega\mathsf{G}_{ij\sigma}(\omega) = \delta_{ij}\mathsf{Q} + \langle\langle [\hat{X}_{i\sigma}, H] \mid \hat{X}^{\dagger}_{j\sigma}\rangle\rangle_{\omega}. \qquad (20)$$

Here the correlation function $\mathsf{Q} = \langle\{\hat{X}_{i\sigma}, \hat{X}^{\dagger}_{i\sigma}\}\rangle = \hat{\tau}_0 \times \hat{Q}$ where $\hat{Q} = \begin{pmatrix} Q_2 & 0 \\ 0 & Q_1 \end{pmatrix}$ and $\hat{\tau}_0$ is the 2×2 unit matrix. The spectral weights of the Hubbard subbands in the paramagnetic state $Q_2 = \langle X_i^{22} + X_i^{\sigma\sigma}\rangle = n/2$ and $Q_1 = \langle X_i^{00} + X_i^{\bar{\sigma}\bar{\sigma}}\rangle = 1 - Q_2$ depend on the average occupation number of holes (6). In the Q matrix we neglect anomalous averages of the type $\langle X_i^{02}\rangle$ which are irrelevant for the d-wave pairing [47].

To introduce the zero-order quasiparticle (QP) excitation energy we use the Mori-type projection operator method [42, 43]. In this approach, the many-particle operator $\hat{Z}_{i\sigma} = [\hat{X}_{i\sigma}, H]$ in (20) is written as a sum of a linear part and an irreducible part $\hat{Z}^{(ir)}_{i\sigma}$ orthogonal to $\hat{X}^{\dagger}_{j\sigma}$:

$$\hat{Z}_{i\sigma} = [\hat{X}_{i\sigma}, H] = \sum_{l} \mathsf{E}_{il\sigma}\hat{X}_{l\sigma} + \hat{Z}^{(\mathrm{ir})}_{i\sigma}. \tag{21}$$

From the orthogonality condition, $\langle\{\hat{Z}^{(\mathrm{ir})}_{i\sigma}, \hat{X}^{\dagger}_{j\sigma}\}\rangle = 0$, we obtain the excitation energy in the generalized mean-field approximation (GMFA) which is given by the time-independent matrix of correlation functions:

$$\mathsf{E}_{ij\sigma} \;=\; \langle\{[\hat{X}_{i\sigma}, H], \hat{X}^{\dagger}_{j\sigma}\}\rangle \mathsf{Q}^{-1}. \tag{22}$$

The corresponding zero-order GF reads

$$\mathsf{G}^0_{\sigma}(\mathbf{k}, \omega) = \Big(\omega\tilde{\tau}_0 - \mathsf{E}_{\sigma}(\mathbf{k})\Big)^{-1} \mathsf{Q}\,, \tag{23}$$

where $\tilde{\tau}_0$ is the 4×4 unit matrix. The spectrum of QP excitations in the GMFA is given by the Fourier component of the matrix (22):

$$\begin{aligned} \mathsf{E}_{\sigma}(\mathbf{k}) &= \frac{1}{N}\sum_{\mathbf{k}} \exp[i\mathbf{k}(\mathbf{i}-\mathbf{j})]\langle\{[\hat{X}_{i\sigma}, H], \hat{X}^{\dagger}_{j\sigma}\}\rangle \mathsf{Q}^{-1} \\ &= \begin{pmatrix} \hat{\varepsilon}(\mathbf{k}) & \hat{\Delta}_{\sigma}(\mathbf{k}) \\ \hat{\Delta}^{*}_{\sigma}(\mathbf{k}) & -\hat{\varepsilon}(\mathbf{k}) \end{pmatrix}, \end{aligned} \tag{24}$$

where $\hat{\varepsilon}(\mathbf{k})$ and $\hat{\Delta}_{\sigma}(\mathbf{k})$ are the normal and anomalous parts of the energy matrix.

To calculate the multiparticle GF $\langle\langle \hat{Z}^{(\mathrm{ir})}_{i\sigma}(t) \mid \hat{X}^{\dagger}_{j\sigma}(t')\rangle\rangle$ in (20) we differentiate it with respect to the second time t' and apply the same projection procedure as in (21). This results in the equation for the GF (20) in the form,

$$\mathsf{G}_{\sigma}(\mathbf{k}, \omega) = \mathsf{G}^0_{\sigma}(\mathbf{k}, \omega) + \mathsf{G}^0_{\sigma}(\mathbf{k}, \omega)\, \mathsf{T}_{\sigma}(\mathbf{k}, \omega)\, \mathsf{G}^0_{\sigma}(\mathbf{k}, \omega), \tag{25}$$

where the scattering matrix

$$\mathsf{T}_{\sigma}(\mathbf{k}, \omega) = \mathsf{Q}^{-1} \langle\langle \hat{Z}^{(\mathrm{ir})}_{\mathbf{k}\sigma} \mid \hat{Z}^{(\mathrm{ir})\dagger}_{\mathbf{k}\sigma}\rangle\rangle_{\omega}\, \mathsf{Q}^{-1}. \tag{26}$$

Now we can introduce the self-energy operator $\Sigma_{\sigma}(\mathbf{q}, \omega)$ as the *proper* part (pp) of the scattering matrix (26) which has no parts connected by the zero-order GF (23) according to the equation: $\mathsf{T} = \Sigma + \Sigma\, \mathsf{G}^0\, \mathsf{T}$. The definition of the proper part of the scattering matrix (26) is equivalent to an introduction of a projected Liouvillian superoperator for the memory function in the conventional Mori technique [42].

Using the self-energy operator instead of the scattering matrix in Eq. (25) we obtain the Dyson equation for the GF (14):

$$\mathsf{G}_\sigma(\mathbf{k},\omega) = [\omega\tilde{\tau}_0 - \mathsf{E}_\sigma(\mathbf{k}) - \mathsf{Q}\Sigma_\sigma(\mathbf{k},\omega)]^{-1}\,\mathsf{Q}, \tag{27}$$

where the self-energy operator is given by

$$\mathsf{Q}\Sigma_\sigma(\mathbf{k},\omega) = \langle\langle \hat{Z}^{(\mathrm{ir})}_{\mathbf{k}\sigma} \mid \hat{Z}^{(\mathrm{ir})\dagger}_{\mathbf{k}\sigma} \rangle\rangle^{(\mathrm{pp})}_{\omega}\,\mathsf{Q}^{-1}. \tag{28}$$

Dyson equation (27) with the zero-order QP excitation energy (24) and the self-energy (28) gives an exact representation for the GF (14). The self-energy takes into account processes of inelastic scattering of electrons (holes) on spin and charge fluctuations due to the kinematical interaction and the dynamic intersite CI and the EPI as shown in Eq. (11).

The self-energy operator (28) can be written in the same matrix form as the GF (17):

$$\mathsf{Q}\Sigma_\sigma(\mathbf{k},\omega) = \begin{pmatrix} \hat{M}_\sigma(\mathbf{k},\omega) & \hat{\Phi}_\sigma(\mathbf{k},\omega) \\ \hat{\Phi}^\dagger_\sigma(\mathbf{k},\omega) & -\hat{M}_{\bar{\sigma}}(\mathbf{k},-\omega) \end{pmatrix}\mathsf{Q}^{-1}, \tag{29}$$

where the matrices $\hat{M}$ and $\hat{\Phi}$ denote the corresponding normal and anomalous (pair) components of the self-energy operator.

The system of equations for the (4×4) matrix GF (17) and the self-energy (29) can be reduced to a system of equations for the normal $\hat{G}_\sigma(\mathbf{k},\omega)$ and the pair $\hat{F}_\sigma(\mathbf{k},\omega)$ matrix components. Using representations for the energy matrix (24) and the self-energy (29), we derive for these components the following system of matrix equations:

$$\hat{G}(\mathbf{k},\omega) = \Big(\hat{G}_N(\mathbf{k},\omega)^{-1} + \hat{\varphi}_\sigma(\mathbf{k},\omega)\,\hat{G}_N(\mathbf{k},-\omega)\,\hat{\varphi}^*_\sigma(\mathbf{k},\omega)\Big)^{-1}\hat{Q}, \tag{30}$$

$$\hat{F}_\sigma(\mathbf{k},\omega) = -\hat{G}_N(\mathbf{k},-\omega)\,\hat{\varphi}_\sigma(\mathbf{k},\omega)\,\hat{G}(\mathbf{k},\omega), \tag{31}$$

where we introduced the normal state GF

$$\hat{G}_N(\mathbf{k},\omega) = \Big(\omega\hat{\tau}_0 - \hat{\varepsilon}(\mathbf{k}) - \hat{M}(\mathbf{k},\omega)/\hat{Q}\Big)^{-1}, \tag{32}$$

and the superconducting gap function

$$\hat{\varphi}_\sigma(\mathbf{k},\omega) = \hat{\Delta}_\sigma(\mathbf{k}) + \hat{\Phi}_\sigma(\mathbf{k},\omega)/\hat{Q}. \tag{33}$$

The system of equations for the normal GFs (30), (32), the anomalous GF (31) and the gap equation (33) should be solved self-consistently for the multiparticle GFs in the self-energy operator (29) as discussed below.

3.2. Generalized Mean-Field Approximation

The superconducting pairing in the Hubbard model already occurs in the GMFA and is caused by the kinetic superexchange interaction as in the t–J model [19]. Therefore, it is reasonable to consider at first the GMFA described by the zero-order GF (23). The energy matrix (22) is calculated using the commutation relations (7) for the HOs. The normal part

$\hat{\varepsilon}(\mathbf{k})$ of the energy matrix (24) after diagonalization determines the QP spectrum in two Hubbard subbands in the GMFA (for detail see [38]):

$$\begin{aligned} \varepsilon_{1,2}(\mathbf{k}) &= (1/2)[\omega_2(\mathbf{k}) + \omega_1(\mathbf{k})] \mp (1/2)\Lambda(\mathbf{k}), \qquad (34) \\ \omega_\iota(\mathbf{k}) &= 4t\,\alpha_\iota \gamma(\mathbf{k}) + 4\,\beta_\iota\, t' \gamma'(\mathbf{k}) + 4\,\beta_\iota\, t'' \gamma''(\mathbf{k}) \\ &+ \omega_\iota^{(c)}(\mathbf{k}) + U\delta_{\iota,2} - \mu, \quad (\iota = 1,2) \qquad (35) \\ \Lambda(\mathbf{k}) &= \{[\omega_2(\mathbf{k}) - \omega_1(\mathbf{k})]^2 + 4W(\mathbf{k})^2\}^{1/2}, \\ W(\mathbf{k}) &= 4t\,\alpha_{12}\gamma(\mathbf{k}) + 4t'\,\beta_{12}\gamma'(\mathbf{k}) + 4t''\,\beta_{12}\gamma''(\mathbf{k}). \end{aligned}$$

Here the hopping parameter is defined by the expression:

$$t_{ij} = (1/N) \sum_{\mathbf{k}} \exp[i\mathbf{k}(\mathbf{i} - \mathbf{j})]\, t(\mathbf{k}), \qquad (36)$$

$$t(\mathbf{k}) = 4t\,\gamma(\mathbf{k}) + 4t'\,\gamma'(\mathbf{k}) + 4t''\,\gamma''(\mathbf{k}). \qquad (37)$$

which are equal to t for the nearest neighbors (nn) sites $a_1 = (\pm a_x, \pm a_y)$, t' – for the next nearest neighbors (nnn) sites $a_d = \pm(a_x \pm a_y)$, and t'' – for the nnn sites $a_2 = (\pm 2a_x, \pm 2a_y)$. The corresponding **k**-dependent functions are: $\gamma(\mathbf{k}) = (1/2)(\cos k_x + \cos k_y)$, $\gamma'(\mathbf{k}) = \cos k_x \cos k_y$, and $\gamma''(\mathbf{k}) = (1/2)(\cos 2k_x + \cos 2k_y)$ (the lattice constants $a_x = a_y$ are put to unity).

The kinematical interaction for the HOs results in renormalization of the spectrum (34) determined by the parameters: $\alpha_\iota = Q_\iota[1 + C_1/Q_\iota^2]$, $\beta_\iota = Q_\iota[1 + C_2/Q_\iota^2]$, $\alpha_{12} = \sqrt{Q_1 Q_2}[1 - C_1/Q_1 Q_2]$, $\beta_{12} = \sqrt{Q_1 Q_2}[1 - C_2/Q_1 Q_2]$. In addition to the conventional Hubbard I renormalization given by Q_1, Q_2 parameters an essential renormalization is caused by the AF spin correlation functions for the nn and the nnn sites, respectively:

$$C_1 = \langle \mathbf{S}_i \mathbf{S}_{i+a_1} \rangle, \quad C_2 = \langle \mathbf{S}_i \mathbf{S}_{i+a_d(a_2)} \rangle. \qquad (38)$$

These functions strongly depend on doping resulting in a considerable variation of the electronic spectrum as shown later and discussed in detail in Ref. [38].

The contribution from the CI V_{ij} in (35) is given by

$$\omega_{1(2)}^{(c)}(\mathbf{k}) = \frac{1}{N} \sum_{\mathbf{q}} V(\mathbf{k} - \mathbf{q}) N_{1(2)}(\mathbf{q}), \qquad (39)$$

where $N_1(\mathbf{q}) = \langle X_{\mathbf{q}}^{0\bar{\sigma}} X_{\mathbf{q}}^{\bar{\sigma}0} \rangle / Q_1$ and $N_2(\mathbf{q}) = \langle X_{\mathbf{q}}^{\sigma 2} X_{\mathbf{q}}^{2\sigma} \rangle / Q_2$ are occupation numbers in the single-particle and two-particle subbands, respectively and $V(\mathbf{q})$ is the Fourier transform of intersite CI V_{ij}.

The anomalous component $\hat{\Delta}_\sigma(\mathbf{k})$ of the matrix (24) determines the superconduction gap in the GMFA. Considering the singlet d-wave pairing, we calculate the intersite pair correlation functions. The diagonal matrix components are given by the equations:

$$\Delta_{ij\sigma}^{22} Q_2 = -\sigma\, t_{ij}^{21} \langle X_i^{02} N_j \rangle - V_{ij} \langle X_i^{\sigma 2} X_j^{\bar{\sigma} 2} \rangle, \qquad (40)$$

$$\Delta_{ij\sigma}^{11} Q_1 = \sigma\, t_{ij}^{12} \langle N_j X_i^{02} \rangle - V_{ij} \langle X_i^{0\bar{\sigma}} X_j^{0\sigma} \rangle. \qquad (41)$$

Here we used the notation t_{ij}^{12} for the interband hopping parameters to emphasize that the kinematical pairing $\langle X_i^{02} N_j \rangle$ is mediated by the interband hopping. In terms of the

Fermi operators $a_{i\sigma} = X_i^{0\sigma} + \sigma X_i^{\bar{\sigma}2}$, the pair correlation function in (40) can be written as $\langle X_i^{02} N_j \rangle = \langle X_i^{0\downarrow} X_i^{\downarrow 2} N_j \rangle = \langle a_{i\downarrow} a_{i\uparrow} N_j \rangle$. This representation shows that the kinematical pairing occurs on a single lattice site but in two Hubbard subbands [37]. The correlation function $\langle X_i^{02} N_j \rangle$ can be calculated directly from the GF $L_{ij}(t-t') = \langle\langle X_i^{02}(t) \mid N_j(t') \rangle\rangle$ without *any decoupling* approximation as shown in Ref. [37]. In particular, under hole doping, $n = 1 + \delta > 1$, the pair correlation function in the two-site approximation reads (for detail see Appendix A):

$$\langle X_i^{02} N_j \rangle = -\frac{4t_{ij}^{12}}{U} \sigma \, \langle X_i^{\sigma 2} X_j^{\bar{\sigma} 2} \rangle. \tag{42}$$

As a result, the equation for the superconducting gap in (40) can be written as

$$\Delta_{ij\sigma}^{22} = (J_{ij} - V_{ij}) \, \langle X_i^{\sigma 2} X_j^{\bar{\sigma} 2} \rangle / Q_2, \tag{43}$$

where $J_{ij} = 4\,(t_{ij}^{12})^2/U$ is the AF superexchange interaction. A similar equation holds for the gap in the one-hole subband: $\Delta_{ij\sigma}^{11} = (J_{ij} - V_{ij}) \, \langle X_i^{0\bar{\sigma}} X_j^{0\sigma} \rangle / Q_1$. We thus conclude that the pairing in the Hubbard model in the GMFA is similar to the superconductivity in the t–J model mediated by the AF superexchange interaction J_{ij} induced by the kinematical interaction for the interband hopping.

3.3. Self-Energy Operator

To determine the self-energy matrix we should calculate many-particle GFs in Eq. (29). Since the diagram technique for HOs is very complicated and can be used effectively only for the lowest order diagrams in the Hubbard model (see, e.g., Refs. [31, 48]) we consider a more simple technique based on the two-time decoupling of multiparticle correlation functions in the projection operator method [43]. In this method the SCBA is used which is equivalent to the noncrossing approximation in the diagram technique. In this approximation, a propagation of Fermi-type excitations described by operators $X_l^{\sigma' 2}$, and Bose-type excitations described by operators $B_{i\sigma\sigma'}$ on different lattice sites, $l \neq i$, is assumed to be independent. Therefore, the time-dependent multiparticle correlation functions in the self-energy operators (29) can be written as a product of fermionic and bosonic correlation functions,

$$\langle X_{l'}^{2\sigma''} B_{j\sigma\sigma''}^{\dagger} | B_{i\sigma\sigma'}(t) X_l^{\sigma' 2}(t) \rangle = \delta_{\sigma',\sigma''} \langle X_{l'}^{2\sigma'} X_l^{\sigma' 2}(t) \rangle \langle B_{j\sigma\sigma'}^{\dagger} | B_{i\sigma\sigma'}(t) \rangle, \tag{44}$$

$$\langle X_{l'}^{\bar{\sigma}'' 2} B_{j\bar{\sigma}\bar{\sigma}''} | B_{i\sigma\sigma'}(t) X_l^{\sigma' 2}(t) \rangle = \delta_{\sigma',\sigma''} \langle X_{l'}^{\bar{\sigma}' 2} X_l^{\sigma' 2}(t) \rangle \, \langle B_{j\bar{\sigma}\bar{\sigma}'} B_{i\sigma\sigma'}(t) \rangle \,. \tag{45}$$

The time-dependent correlation functions are calculated self-consistently using the corresponding GFs (for detail see Appendix B).

In particular, the normal and anomalous diagonal components of the self-energy for the two-particle subband are determined by the expressions

$$\begin{aligned} M^{22}(\mathbf{k}, \omega) &= \frac{1}{N} \sum_{\mathbf{q}} \int_{-\infty}^{+\infty} \mathrm{d}z \, K^{(+)}(\omega, z | \mathbf{q}, \mathbf{k} - \mathbf{q}) \\ &\times \Big\{ -(1/\pi) \mathrm{Im} \big[G^{22}(\mathbf{q}, z) + G^{11}(\mathbf{q}, z) \big] \Big\}, \end{aligned} \tag{46}$$

$$\Phi^{22}_{\sigma}(\mathbf{k},\omega) = \frac{1}{N}\sum_{\mathbf{q}}\int_{-\infty}^{+\infty} \mathrm{d}z\, K^{(-)}(\omega, z|\mathbf{q},\mathbf{k}-\mathbf{q}) \times \Big\{-(1/\pi)\mathrm{Im}\left[F^{22}_{\sigma}(\mathbf{q},z) - F^{11}_{\sigma}(\mathbf{q},z)\right]\Big\}, \tag{47}$$

where $G^{\alpha\alpha}(\mathbf{q}, z)$ and $F^{\alpha\alpha}_{\sigma}(\mathbf{q}, z)$ are given by the diagonal components of the matrices (30), (31). Similar expressions hold for the self-energy components $M^{11}(\mathbf{k},\omega)$ and $\Phi^{11}_{\sigma}(\mathbf{k},\omega)$ for the one-particle subband (see Ref. [36]). Note, that in the paramagnetic normal state the GF (30) and the self-energy (46) do not depend on the spin σ.

The kernel of the integral equations (46), (47) has a form, similar to the strong-coupling Migdal-Eliashberg theory [49, 50]:

$$K^{(\pm)}(\omega, z|\mathbf{q},\mathbf{k}-\mathbf{q}) = \frac{1}{\pi}\int_{-\infty}^{+\infty} \mathrm{d}\Omega\, \frac{1+N(\Omega)-n(z)}{\omega - z - \Omega} \times\Big\{|t(\mathbf{q})|^2 \mathrm{Im}\chi_{sf}(\mathbf{k}-\mathbf{q},\Omega) \pm |g(\mathbf{k}-\mathbf{q})|^2 \mathrm{Im}\chi_{ph}(\mathbf{k}-\mathbf{q},\Omega) \pm \left[|V(\mathbf{k}-\mathbf{q})|^2 + |t(\mathbf{q})|^2/4\right] \mathrm{Im}\,\chi_{cf}(\mathbf{k}-\mathbf{q},\Omega)\Big\}, \tag{48}$$

where $n(\omega) = [e^{\omega/T}+1]^{-1}$ and $N(\omega) = [e^{\omega/T}-1]^{-1}$. The spectral densities of bosonic excitations are determined by the dynamic susceptibility for spin (sf), number (charge) (cf), and lattice (phonon) (ph) fluctuations

$$\chi_{sf}(\mathbf{q},\omega) = -\langle\langle \mathbf{S}_{\mathbf{q}}|\mathbf{S}_{-\mathbf{q}}\rangle\rangle_{\omega}, \tag{49}$$

$$\chi_{cf}(\mathbf{q},\omega) = -\langle\langle \delta N_{\mathbf{q}}|\delta N_{-\mathbf{q}}\rangle\rangle_{\omega}, \tag{50}$$

$$\chi_{ph}(\mathbf{q},\omega) = -\langle\langle u_{\mathbf{q}}|u_{-\mathbf{q}}\rangle\rangle_{\omega}, \tag{51}$$

which are defined in terms of the commutator GFs [44] for the spin $\mathbf{S}_{\mathbf{q}}$, number $\delta N_{\mathbf{q}} = N_{\mathbf{q}} - \langle N_{\mathbf{q}}\rangle$, and lattice displacement (phonon) $u_{\mathbf{q}}$ operators.

In the SCBA (44), (45), vertex corrections to the kinematical interaction $t(\mathbf{q})$ of electrons with spin- and charge-fluctuations (49), (50) induced by the intraband hopping are neglected. It is assumed that the system is far away from a charge instability or a stripe formation and charge-fluctuations give a small contribution to the pairing. The largest contribution from spin fluctuations comes from wave vectors close the AF wave vector $\mathbf{Q} = (\pi,\pi)$ where their energy $\omega_s(\mathbf{Q})$ is much smaller than the Fermi energy, $\omega_s(\mathbf{Q})/E_{\mathrm{F}} \ll 1$ (see, e.g., [51]). Therefore, vertex corrections to the kinematical interaction should be small as in Eliashberg theory [50] for electron interaction with phonons, where $\omega_{ph}/E_{\mathrm{F}} \ll 1$. Consequently, the SCBA for the self-energy and the GFs calculated self-consistently is quite reliable and makes it possible to examine the strong coupling regime which is essential in study of renormalization of the QP spectrum and the superconducting pairing as shown in Refs. [38, 39] and discussed later.

4. Results and Discussion

4.1. Self-Consistent System of Equation

The self-consistent system of equations for the diagonal components of normal GFs (30), (32) and the self-energy (46) can be written in the form [38]:

$$G_N^{11(22)}(\mathbf{k},\omega) = [1-b(\mathbf{k})]G_{1(2)}(\mathbf{k},\omega)+b(\mathbf{k})G_{2(1)}(\mathbf{k},\omega), \tag{52}$$

where the hybridization parameter $b(\mathbf{k}) = [\varepsilon_2(\mathbf{k}) - \omega_2(\mathbf{k})]/[\varepsilon_2(\mathbf{k}) - \varepsilon_1(\mathbf{k})]$. The two subband GFs in Eq. (52) in the imaginary frequency representation, $i\omega_n = i\pi T(2n+1)$, $n = 0, \pm1, \pm2, \ldots$, take the form:

$$\{G_{1(2)}(\mathbf{k},\omega_n)\}^{-1} = i\omega_n - \varepsilon_{1(2)}(\mathbf{k}) - \Sigma(\mathbf{k},\omega_n). \tag{53}$$

The self-energy $\Sigma(\mathbf{k},\omega)$ can be approximated by the same function for two subbands:

$$\begin{aligned}\Sigma(\mathbf{k},\omega_n) &= -\frac{T}{N}\sum_{\mathbf{q}}\sum_{m}\lambda^{(+)}(\mathbf{q},\mathbf{k}-\mathbf{q}\mid\omega_n-\omega_m)[G_1(\mathbf{q},\omega_m)+G_2(\mathbf{q},\omega_m)]\\ &\equiv i\omega_n\,[1-Z(\mathbf{k},\omega_n)]+X(\mathbf{k},\omega_n).\end{aligned} \tag{54}$$

Density of states (DOS) is determined by

$$A(\omega) = \frac{1}{N}\sum_{\mathbf{k}} A(\mathbf{k},\omega), \tag{55}$$

where the spectral function reads

$$\begin{aligned}A(\mathbf{k},\omega) &= [Q_1+P(\mathbf{k})]A_1(\mathbf{k},\omega)+[Q_2-P(\mathbf{k})]A_2(\mathbf{k},\omega),\\ A_{1(2)}(\mathbf{k},\omega) &= -(1/\pi)\,\mathrm{Im}G_{1(2)}(\mathbf{k},\omega).\end{aligned} \tag{56}$$

Here the hybridization parameter $P(\mathbf{k}) = (n-1)b(\mathbf{k}) - 2\sqrt{Q_1Q_2}\,[W(\mathbf{k})/\Lambda(\mathbf{k})]$ takes into account contributions from both the diagonal and off-diagonal components of the GF (32).

To calculate T_c we can use a linear approximation for the pair GF (31). In particular, Eq. (33) for the two-particle subband gap $\varphi(\mathbf{k},\omega) = \sigma\varphi_{2,\sigma}(\mathbf{k},\omega)$ can be written as

$$\begin{aligned}\varphi(\mathbf{k},\omega_n) &= \frac{T_c}{N}\sum_{\mathbf{q}}\sum_{m}\frac{[1-b(\mathbf{q})]^2\,\varphi(\mathbf{q},\omega_m)}{[\omega_m Z(\mathbf{q},\omega_m)]^2+[\varepsilon_2(\mathbf{q})+X\mathbf{q},\omega_m)]^2}\\ &\times\ \{\,J(\mathbf{k}-\mathbf{q})-V(\mathbf{k}-\mathbf{q})+\lambda^{(-)}(\mathbf{q},\mathbf{k}-\mathbf{q}\mid\omega_n-\omega_m)\}.\end{aligned} \tag{57}$$

The interaction functions in (54) and (57) are given by

$$\begin{aligned}\lambda^{(\pm)}(\mathbf{q},\mathbf{k}-\mathbf{q}|\nu_n) &= \mp\Big\{\,[|V(\mathbf{k}-\mathbf{q})|^2+|t(\mathbf{q})|^2/4]\,\chi_{cf}(\mathbf{k}-\mathbf{q},\nu_n)\\ &+|g(\mathbf{k}-\mathbf{q})|^2\,\chi_{ph}(\mathbf{k}-\mathbf{q},\nu_n)\Big\} - |t(\mathbf{q})|^2\,\chi_{sf}(\mathbf{k}-\mathbf{q},\nu_n).\end{aligned} \tag{58}$$

Thus, we have derived the self-consistent system of equations for the normal GF (53), the self-energy (54), and the gap function (57).

In the present theory we do not perform self-consistent computation of spin and charge excitation spectra but adopt certain models for the spin (49), charge (50), and phonon (51) susceptibility in Eq. (48) or Eq. (58). Since we consider the electronic spectrum only in the normal state and calculate superconducting transition temperature T_c from the linearized gap equation (57) the feedback effects caused by opening a superconducting gap are not essential which justifies usage of model functions for the susceptibility. Due to a large energy scale of charge fluctuations, of the order of several t, in comparison with the spin excitation energy of the order of J, the charge fluctuation contributions can be considered in the static limit for the susceptibility (50)

$$\begin{aligned} \chi_{cf}(\mathbf{k}) &= \chi_{cf}^{(1)}(\mathbf{k}) + \chi_{cf}^{(2)}(\mathbf{k}), \\ \chi_{cf}^{(\alpha)}(\mathbf{k}) &= -\frac{1}{N}\sum_{\mathbf{q}} \frac{N^{(\alpha)}(\mathbf{q}+\mathbf{k}) - N^{(\alpha)}(\mathbf{q})}{\varepsilon_\alpha(\mathbf{q}+\mathbf{k}) - \varepsilon_\alpha(\mathbf{q})}, \end{aligned} \tag{59}$$

where the occupation numbers $N^{(\alpha)}(\mathbf{q})$ are defined as

$$\begin{aligned} N^{(1)}(\mathbf{k}) &= [Q_1 + (n-1)b(\mathbf{k})]\, N_1(\mathbf{k}), \\ N^{(2)}(\mathbf{k}) &= [Q_2 - (n-1)b(\mathbf{k})]\, N_2(\mathbf{k}), \\ N_\alpha(\mathbf{k}) &= (1/2) + T\sum_m G_\alpha(\mathbf{k}, \omega_m). \end{aligned} \tag{60}$$

For the dynamical spin susceptibility $\chi_{sf}(\mathbf{q},\omega)$ (49) we used a model suggested in Ref. [52]

$$\begin{aligned} &\operatorname{Im}\chi_{sf}(\mathbf{q}, \omega + i0^+) = \chi_{sf}(\mathbf{q})\, \chi_{sf}''(\omega) \\ &= \frac{\chi_Q}{1+\xi^2[1+\gamma(\mathbf{q})]} \tanh\frac{\omega}{2T}\frac{1}{1+(\omega/\omega_s)^2}\,. \end{aligned} \tag{61}$$

Such type of the spin-excitation spectrum was found in the microscopic theory for the t-J model in Ref. [51]. The model is determined by two parameters: the AF correlation length ξ and the cut-off energy of spin excitations of the order of the exchange energy $\omega_s \sim J$. The strength of the spin-fluctuation interaction given by the static susceptibility $\chi_Q = \chi_{sf}(\mathbf{Q})$

$$\chi_Q = \frac{3(1-\delta)}{2\omega_s}\left\{\frac{1}{N}\sum_{\mathbf{q}} \frac{1}{1+\xi^2[1+\gamma(\mathbf{q})]}\right\}^{-1}, \tag{62}$$

is defined by the normalization condition:

$$\frac{1}{N}\sum_{\mathbf{q}} \int_0^\infty \frac{d\omega}{\pi}\coth\frac{\omega}{2T}\operatorname{Im}\chi_{sf}(\mathbf{q},\omega) = \langle \mathbf{S}_i^2\rangle = \frac{3}{4}(1-\delta).$$

The spin correlation functions (38) in the single-particle excitation spectrum (34) are calculated using the same model (61): $C_1 = (1/N)\sum_{\mathbf{q}} C_{\mathbf{q}}\,\gamma(\mathbf{q})$, $C_2 = (1/N)\sum_{\mathbf{q}} C_{\mathbf{q}}\,\gamma'(\mathbf{q})$, where $C_{\mathbf{q}} = (\omega_s/2)(\chi_Q/\{1+\xi^2[1+\gamma(\mathbf{q})]\})$.

To estimate the contribution from phonons in Eq. (48) we consider a model susceptibility for optic phonons and the EPI matrix element in the form similar to Ref. [53]:

$$V_{ep}(\mathbf{q},\omega) = |g(\mathbf{q})|^2 \chi_{ph}(\mathbf{q},\omega) = g_{ep}\,\frac{\omega_0^2}{\omega_0^2-\omega^2}\,S(q), \tag{63}$$

where g_{ep} is the "bare" matrix element for the short-range EPI, while the momentum dependence of the EPI is determined by the vertex correction $S(q)$. It takes into account a strong suppression of charge fluctuations at small distances (large scattering momenta q) induced by electron correlations as proposed in Ref. [41]. For the vertex function we take the model

$$S(q) = \frac{1}{\kappa_1^2+q^2} \equiv \frac{\xi_{ch}^2}{1+\xi_{ch}^2\,q^2}, \tag{64}$$

where the charge correlation length $\xi_{ch} = 1/\kappa_1$ determines the radius of a "correlation hole". Taking into account that $\xi_{ch} \sim a/\delta$ [41], we can use the relation $\xi_{ch} = 1/(2\delta)$ in numerical computations. This gives $\xi_{ch} \simeq 10$ for the underdoped case ($\delta = 0.05$) and $\xi_{ch} \simeq 2$ for the overdoped case ($\delta = 0.25$). We assume a strong EPI $g_{ep} = 5\,t$ and take $\omega_0 = 0.1\,t$.

For the intersite CI V_{ij} we consider a model for repulsion of two electrons (holes) on neighbor lattice sites,

$$V(\mathbf{q}) = 2V\,(\cos q_x + \cos q_y)\,, \tag{65}$$

with various values of $V = 0.0,\ 0.5\,t,\ 1.0\,t$ and $2.0\,t$. Most of the calculations are done for $U = 8\,t$ while several results are presented for $U = 4\,t,\ 16\,t$ and $32\,t$. The AF exchange interaction for neighbor sites is described by the function $J(\mathbf{q}) = 2J\,(\cos q_x + \cos q_y)$ with $J = 0.4t$. Note, that in the GMFA the CI V_{ij} gives no contribution to the exchange interaction J_{ij} and therefore it is assumed to be the same for all values of V (cf. with Refs. [54, 55]). In computations we use the following hoping parameters $t' = -0.2\,t,\quad t'' = 0.10\,t$ where $t = 0.4$ eV is the energy unit.

4.2. Electronic Spectrum in the Normal State

At first we consider results in the GMFA for the electronic spectrum (34). The doping dependence of the electron dispersion for the two-hole subband $\varepsilon_2(\mathbf{k})$ along the symmetry directions in the 2D Brillouin zone (BZ) are shown in Fig. 1 for $U = 8$ for $V = 0$ (a) and for $V = 2$ (b). The corresponding Fermi surfaces (FSs) determined by the equation: $\varepsilon_2(\mathbf{k}_{\rm F}) = 0$ are plotted in Fig. 2. For small doping, $\delta = 0.05$, the energy at the $M(\pi,\pi)$ and $\Gamma(0,0)$ points are nearly equal as in the AF phase. Only small hole-like FS pockets close to the $(\pm\pi/2, \pm\pi/2)$ points emerge at this doping as shown in Figs. 1, 2. With increasing doping, the AF correlation length decreases that results in increasing of the electron energy at the $M(\pi,\pi)$ point and at some critical doping $\delta \sim 0.12$ a large FS emerges. The doping dependence of the AF correlation length ξ on several values of hole concentrations δ are given in Table 1. A remarkable feature is that the part of the FS close to the $\Gamma(0,0)$ point in the nodal direction in Fig. 2 does not shift much with doping (or temperature) being pinned to a large FS as observed in ARPES experiments (see, e.g. [56]). At the same time, the renormalized two-hole subband width increases with doping, as e.g., for $U = 8$ and $V = 0$

from $\widetilde{W} \approx 2t$ at $\delta = 0.05$ to $\widetilde{W} \approx 3t$ at $\delta = 0.25$, which, however, remains less than the "bare" Hubbard subband width $W = 4t\,(1+\delta)$ where short-range AF correlations are disregarded. Note that in the dynamical mean field theory (DMFT) this narrowing of the subbands due to the short-range AF correlations is missed [57, 58], while they are partly taken into account in the cluster DMFT [59]. With increasing V the subband width shrinks as seen from comparison panels (a) and (b) for electronic spectra in Figs. 1 and the FS in Figs. 2.

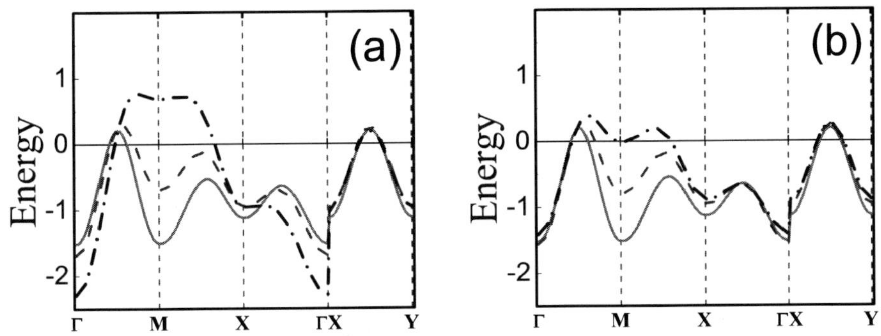

Figure 1. (Color online) Electron dispersion in the GMFA $\varepsilon_2(\mathbf{k})$ for (a) $V = 0$ and (b) $V = 2$ at $U = 8$ along the symmetry directions $\Gamma(0,0) \to M(\pi,\pi) \to X(\pi,0) \to \Gamma(0,0)$ and $X(\pi,0) \to Y(0,\pi)$ for $\delta = 0.05$ (red solid line), 0.10 (blue dashed line), and 0.25 (black dash-dotted line). Fermi energy for hole doping is at $\omega = 0$.

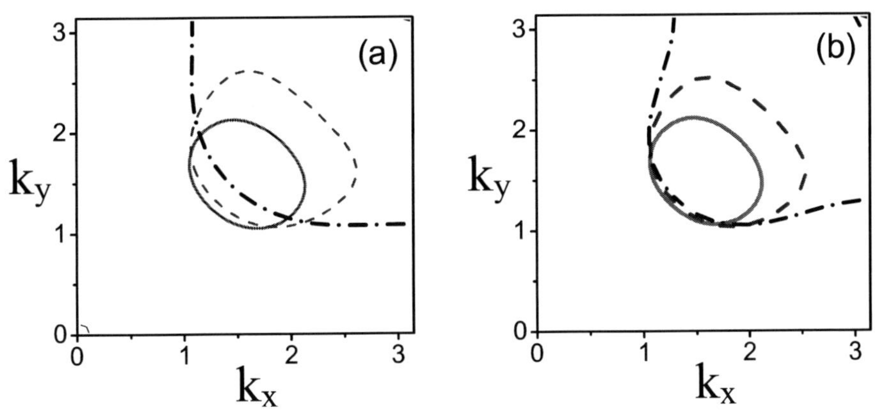

Figure 2. (Color online) Fermi surface for (a) $V = 0$ and (b) $V = 2$ at $U = 8$ in the quarter of the BZ in the GMFA at hole doping $\delta = 0.05$ (red solid line), 0.10 (blue dashed line), and 0.25 (black dash-dotted line).

To consider the self-energy effects in the electronic spectrum a strong coupling approximation (SCA) should be considered by a self-consistent solution of the system of equations for the normal GF (53) and the self-energy (54). In Ref. [38] a detailed investigation of the normal state electronic spectrum for the conventional Hubbard model in SCA was performed. Therefore, here we present only the results of the electronic spectrum computation for the model (3) which are important for further studies of superconductivity in the model.

The spectral functions (56) along the symmetry directions are presented in the left panels in Figs. 3 and 4 for $\delta = 0.10$ and $\delta = 0.25$, respectively. The dispersion curves given by the maximum of the spectral function (56) at the same doping are displayed in the right panels in Figs. 3 and 4.

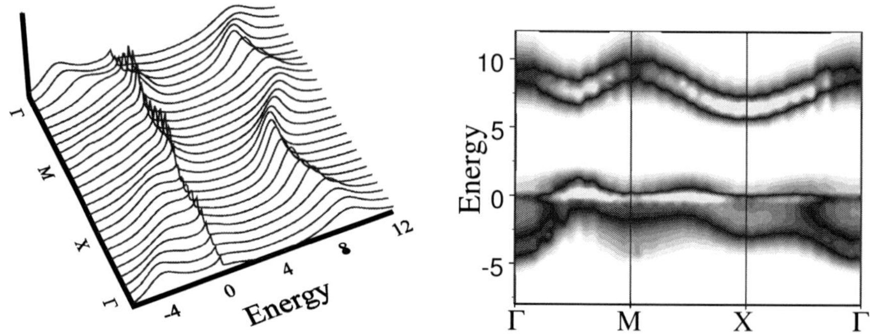

Figure 3. Spectral function (left panel) and electron dispersion curves (right panel) in the SCA along the symmetry directions $\Gamma(0,0) \to M(\pi,\pi) \to X(\pi,0) \to \Gamma(0,0)$ for hole concentration $\delta = 0.10$.

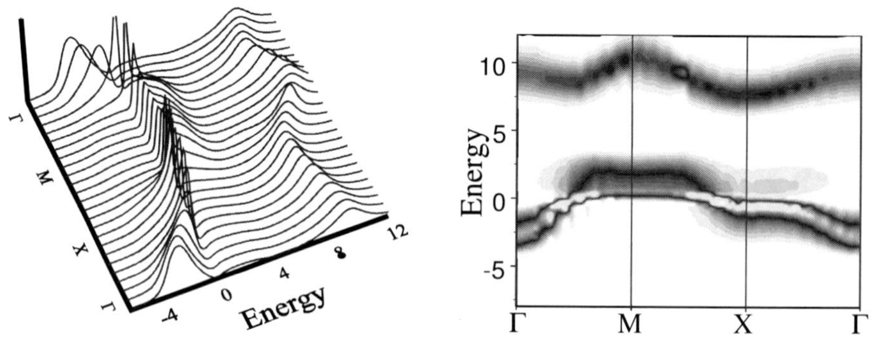

Figure 4. Spectral function (left panel) and electron dispersion curves (right panel) in the SCA along the symmetry directions for hole concentration $\delta = 0.25$.

In comparison with the GMFA in Fig. 1, a rather flat energy dispersion is found with QP peaks at the FS. In general, strong increase of the dispersion and intensity of the QP peaks is observed in the overdoped region in comparison with the underdoped region. This is in agreement with our detailed studies of temperature and doping dependence of the self-energy (54) and spectral function (56) in [38] which have proved strong influence of AF spin-correlations on the spectra.

To estimate the coupling constant $\lambda(\mathbf{q})$ in the two-hole subband, we calculated the renormalization parameter $Z(\mathbf{q})$ in Eq. (54) at the Fermi energy,

$$\begin{aligned} Z(\mathbf{q}) &= Z(\mathbf{q}, \omega = 0) = 1 + \lambda(\mathbf{q}) \\ &= 1 - [\, d\,\mathrm{Re}\,\Sigma(\mathbf{q}, \omega)/d\omega]|_{\omega=0}. \end{aligned} \tag{66}$$

The doping dependence of $Z(\mathbf{q})$ is shown in the left panel in Fig. 5. It weakly depends on δ in the underdoped case for $\delta \lesssim 0.15$ but sharply decreases in the overdoped case for

$\delta \gtrsim 0.25$. The temperature dependence of $Z(\mathbf{q})$ presented in the right panel in Fig. 5 is weak at temperatures lower than the characteristic energy of spin fluctuations $\omega_s \sim J$. The EPI gives a small contribution to the coupling constant as follows from the comparison of $Z(\mathbf{q})$ induced by both spin-fluctuations and EPI contributions (red solid line) with the contribution caused only by spin-fluctuations (blue dashed line).

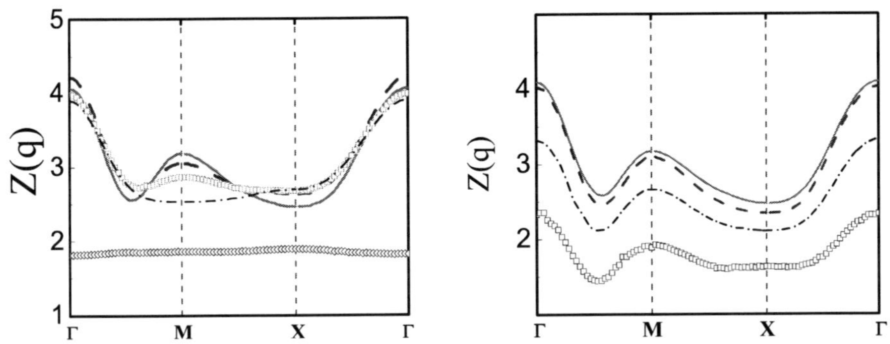

Figure 5. (Color online) Left panel: Doping dependence of the renormalization parameter $Z(\mathbf{q})$ along the symmetry directions $\Gamma(0,0) \rightarrow M(\pi,\pi) \rightarrow X(\pi,0) \rightarrow \Gamma(0,0)$ at $T \approx$ 140 K for $\delta = 0.05$ (red solid line), $\delta = 0.10$ (blue dashed line), $\delta = 0.15$ (red squares), $\delta =$ 0.25 (black dash-dotted line), and $\delta = 0.35$ (black diamonds). Right panel: Temperature dependence of the renormalization parameter $Z(\mathbf{q})$ for $\delta = 0.05$ at $T \approx 140$ K (red solid line), $T \approx 580$ K (black dash-dotted line), and $T \approx 1100$ K (black squares). Blue dashed line shows $Z(\mathbf{q})$ for $\delta = 0.05$ caused only by the spin-fluctuation contribution.

Noticeable changes are observed for the FS in the SCA shown in Fig. 6 which is determined by the equation for the spectral density (56) at the Fermi energy, $A(\mathbf{k}, \omega = 0)$. Whereas in GMFA at small doping the FS in Fig. 2 shows closed pockets, in SCA the FS reveals arc-type behavior detected in ARPES experiments where spectral function $A(\mathbf{k}, \omega = 0)$ is measured (see, e.g., Ref. [60]).

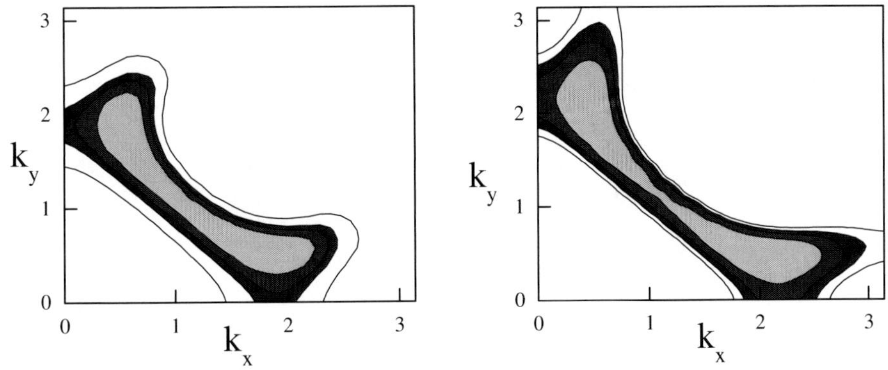

Figure 6. (Color online) $A(\mathbf{k}, \omega = 0)$ on the FS at $\delta = 0.05$ (left panel) and $\delta = 0.1$ (right panel) at $T = 0.03t$ for $U = 4t$.

Thus, studies of the normal state electronic spectrum reveal a major role of AF correlations and self-energy effects in the renormalization of the spectrum. The latter show a noticeable reduction of the QP weight at low doping that strongly suppresses superconducting pairing as we demonstrate below.

4.3. Superconducting Gap and T_c

For a comparison of various contributions to the superconducting gap equation (57), we approximate the interaction (58) by its value close to the Fermi energy. As the result instead of the dynamical susceptibility (49), (50) the static susceptibility $\chi(\mathbf{q}) = \mathrm{Re}\,\chi(\mathbf{q}, \Omega = 0)$ appears in the gap equation. It brings us to the BCS-type equation for the gap function (57) at the Fermi energy $\varphi(\mathbf{k}) = \varphi(\mathbf{k}, \omega = 0)$:

$$\begin{aligned}\varphi(\mathbf{k}) = &\frac{1}{N}\sum_{\mathbf{q}} \frac{[1-b(\mathbf{q})]^2\,\varphi(\mathbf{q})}{[Z(\mathbf{q})]^2\,2\widetilde{\varepsilon}(\mathbf{q})} \tanh\frac{\widetilde{\varepsilon}(\mathbf{q})}{2T_c}\{J(\mathbf{k}-\mathbf{q}) \\ &-V(\mathbf{k}-\mathbf{q}) + [(1/4)|t(\mathbf{q})|^2 + |V(\mathbf{k}-\mathbf{q})|^2]\chi_{cf}(\mathbf{k}-\mathbf{q}) \\ &+|g(\mathbf{k}-\mathbf{q})|^2\,\chi_{ph}(\mathbf{k}-\mathbf{q})\,\theta(\omega_0 - |\varepsilon_2(\mathbf{q})|) \\ &-|t(\mathbf{q})|^2\,\chi_{sf}(\mathbf{k}-\mathbf{q})\theta(\omega_s - |\varepsilon_2(\mathbf{q})|)\}\,, \end{aligned} \tag{67}$$

where $\widetilde{\varepsilon}(\mathbf{q}) = \varepsilon_2(\mathbf{q})/Z(\mathbf{q})$ is the renormalized energy. Whereas for the exchange interaction and CI there are no retardation effects and the pairing occurs for all electrons in the two-particle subband, the EPI and spin-fluctuation contributions are restricted to the range of energies $\pm\omega_0$ and $\pm\omega_s$, respectively, near the FS as determined by the θ-function.

To estimate various contributions in the gap equation (67) we consider a model d-wave gap function, $\varphi(\mathbf{k}) = (\Delta/2)\,\eta(\mathbf{k})$ where $\eta(\mathbf{k}) = (\cos k_x - \cos k_y)$. Then the gap equation can be written in the form (for detail see Ref. [39]):

$$\begin{aligned} 1 \;=\; &\frac{1}{N}\sum_{\mathbf{q}} \frac{[1-b(\mathbf{q})]^2\,[\eta(\mathbf{q})]^2}{[Z(\mathbf{q})]^2\,2\widetilde{\varepsilon}(\mathbf{q})} \tanh\frac{\widetilde{\varepsilon}(\mathbf{q})}{2T_c}\{J - V + (1/4)\,|t(\mathbf{q})|^2\widehat{\chi}_{cf} \\ + \;&\widehat{V}_{cf} + \widehat{V}_{ep}\,\theta(\omega_0 - |\varepsilon_2(\mathbf{q})|) - |t(\mathbf{q})|^2\,\widehat{\chi}_{sf}\theta(\omega_s - |\widetilde{\varepsilon}(\mathbf{q})|)\}. \end{aligned} \tag{68}$$

In this equation only $l = 2$ components of the static susceptibility and CI give contributions

$$\widehat{V}_{cf} = \frac{1}{N}\sum_{\mathbf{k}} |V(\mathbf{k})|^2\,\chi_{cf}(\mathbf{k})\,\cos k_x, \tag{69}$$

$$\widehat{\chi}_{cf} = \frac{1}{N}\sum_{\mathbf{k}} \chi_{cf}(\mathbf{k})\cos k_x, \tag{70}$$

$$\widehat{V}_{ep} = \frac{g_{ep}}{N}\sum_{\mathbf{k}} S(\mathbf{k})\cos k_x, \tag{71}$$

$$\widehat{\chi}_{sf} = \frac{1}{N}\sum_{\mathbf{k}} \chi_{sf}(\mathbf{k})\cos k_x\,. \tag{72}$$

The contribution from the charge fluctuations $\widehat{\chi}_{cf}$ (70) weakly depends on U and V and is very small: $\widehat{\chi}_{cf} \sim 10^{-3}\,(1/t) - 10^{-2}\,(1/t)$ for hole concentrations $\delta = 0.05 - 0.10$,

Table 1. AF correlation length ξ, spin correlation functions C_1, C_2, spin susceptibility χ_Q, projected spin susceptibility $\widehat{\chi}_{sf}$, and EPI parameter $\widehat{V_{ep}}$ for several values of hole concentration δ

δ	ξ/a	C_1	C_2	$\chi_Q \cdot t$	$-\widehat{\chi}_{sf} \cdot t$	$\widehat{V}_{ep}/t$
0.05	3.4	- 0.26	0.16	29.5	1.32	1.96
0.10	2.4	- 0.20	0.11	12.6	1.05	1.4
0.25	1.5	- 0.12	0.05	6.8	0.61	0.76

respectively. For the averaged over the BZ vertex $\overline{|t(\mathbf{q})|^2} = (1/N)\sum_{\mathbf{q}} |t(\mathbf{q})|^2 \simeq 4\,t^2$ the contribution induced by the kinematical interaction is equal to $\overline{|t(\mathbf{q})|^2}\,\widehat{\chi}_{cf} \lesssim 0.04\,t$ and can be neglected. The charge fluctuation contribution $\widehat{V}_{cf}$ (69) from the intersite CI (65) for the hole concentration $\delta = 0.05$ is also small, $\widehat{V}_{cf} \lesssim 5 \cdot 10^{-2}\,t$ and $\widehat{V}_{cf} - V < 0$ for all values of U and V and consequently, the d-wave pairing induced only by charge fluctuations cannot occur.

The EPI contribution (71) for the EPI model (63) even for a strong EPI coupling $g_{ep} = 5t$ is quite small for the d-wave pairing, $\widehat{V}_{ep} = 2\,t\,(0.8\,t)$ for $\delta = 0.05\,(0.25)$, as shown in Table 1. The EPI contribution to the s-wave pairing is given by the $l = 0$ component $S_0 = (1/N)\sum_{\mathbf{q}} S(q) = 0.31\ (0.57)$ for $\xi_{ch} = 2\ (10)$, respectively. The ratio of the d-wave S_d and the s-wave S_0 components of the EPI matrix elements is equal to $(S_d/S_0) = 0.43\ (0.60)$ for $\xi_{ch} = 2\ (10)$, respectively. This shows that at small hole concentrations δ (large charge correlation lengths $\xi_{ch} = 1/2\delta$) the EPI for the both components are comparable, while for the overdoped case the d-wave component S_d becomes considerably smaller than the s-wave component in agreement with the results of Ref. [41].

The spin-fluctuation contribution $\widehat{\chi}_{sf}$ (72) is calculated for the model $\chi_{sf}(\mathbf{q})$ in Eq. (61). Since the spin susceptibility has a maximum at the AF wave vector $\mathbf{Q} = (\pi, \pi)$ the integral over $\mathbf{k}$ in (72) results in the negative value for $\widehat{\chi}_{sf}$ which strongly depends on hole doping as shown in Table 1. Using the averaged over BZ vertex $\overline{|t(\mathbf{q})|^2} \simeq 4\,t^2$ we can estimate an effective spin-fluctuation coupling constant as $g_{sf} \simeq -4\,t^2\,\widehat{\chi}_{sf} = 5\,t - 2\,t$ for $\delta = 0.05 - 0.25$. Thus, the spin-fluctuation contribution to the pairing in Eq. (68) with the coupling constant g_{sf} appears to be the largest.

At first we calculate doping dependence of T_c in the WCA for $Z(\mathbf{q}) = 1$ in Eq. (68) by taking into account the exchange interaction J, the Coulomb repulsion $\widehat{V}_c$, and the contributions from the self-energy $\widehat{V}_{ep}$, $\widehat{\chi}_{sf}$, and $\widehat{V}_{cf}$, neglecting the small contribution $\widehat{\chi}_{cf}$. Results of the calculation is shown in the left panel in Fig. 7. The highest $T_c \approx 0.22\,t$ is found when all the contributions are taken into account. The spin-fluctuation pairing results in superconducting $T_c^{sf} \approx 0.1\,t$ much larger than $T_c^{ep} \approx 0.012t$ mediated by the EPI. For the $\mathbf{k}$-independent EPI $(S(\mathbf{k}) = 1)$ the d-wave pairing is absent. The doping dependence of T_c is qualitatively agree with experiments in cuprates but its value is an order of magnitude higher.

The high values for T_c found in the WCA are explained by neglecting the reduction

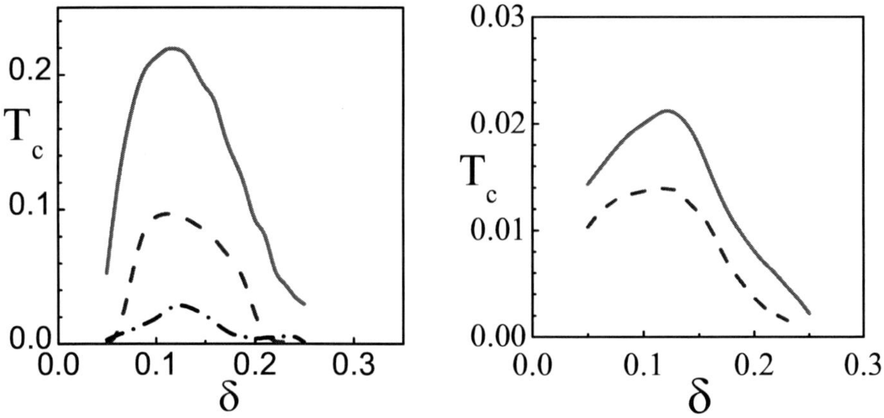

Figure 7. (Color online) $T_c(\delta)$ in the WCA (left panel) and in the SCA (right panel) induced by all interactions (red solid line) and only by the spin-fluctuation contribution (blue dashed line) or only by the EPI $\widehat{V}_{ep}$ on the left panel (black dash-dotted line).

of the QP weight caused by the self-energy effects in the gap equation (57). In the strong coupling approximation (SCA) we should solve the gap equation (57) which is convenient to write in the form:

$$\varphi(\mathbf{k}, \omega_n) = \frac{T_c}{N} \sum_{\mathbf{q}} \sum_{m} \frac{[1 - b(\mathbf{q})]^2 \, \varphi(\mathbf{q}, \omega_m)}{[\omega_m Z(\mathbf{q}, \omega_m)]^2 + [\varepsilon_2(\mathbf{q}) + X(\mathbf{q}, \omega_m)]^2} \tag{73}$$
$$\times \Big\{ J(\mathbf{k} - \mathbf{q}) - V(\mathbf{k} - \mathbf{q}) - V_{sf}(\mathbf{q}, \mathbf{k} - \mathbf{q}, \omega_n - \omega_m) + V_{ep}(\mathbf{k} - \mathbf{q}, \omega_n - \omega_m) \Big\}.$$

For $V(\mathbf{k} - \mathbf{q})$ we take the nearest-neighbor CI (65). Since the charge-fluctuations gives a much weaker contribution than the spin-fluctuation and electron-phonon interactions (see Table 1), we neglect the term $[|V(\mathbf{k} - \mathbf{q})|^2 + |t(\mathbf{q})|^2/4] \, \chi_{cf}(\mathbf{k} - \mathbf{q}, \nu_n)$ in the interaction function (58). Contributions induced by spin-fluctuations and the EPI are described by the functions

$$V_{sf}(\mathbf{q}, \mathbf{k} - \mathbf{q}, \omega_\nu) = |t(\mathbf{q})|^2 \, \chi_{sf}(\mathbf{k} - \mathbf{q}) F_{sf}(\omega_\nu), \tag{74}$$

$$V_{ep}(\mathbf{k} - \mathbf{q}, \omega_\nu) = g_{ep} \, \frac{\omega_0^2}{\omega_0^2 + \omega_\nu^2} \, S(\mathbf{k} - \mathbf{q}), \tag{75}$$

where the spectral function for spin fluctuations reads:

$$F_{sf}(\omega_\nu) = \frac{1}{\pi} \int_0^\infty \frac{2x dx}{x^2 + (\omega_\nu/\omega_s)^2} \, \frac{\tanh(x \, \omega_s/2T)}{1 + x^2} \, . \tag{76}$$

To calculate T_c and to find out the energy- and $\mathbf{k}$-dependence of the gap $\varphi(\mathbf{k}, \omega)$, Eq. (73) was solved by a direct diagonalization in $(\mathbf{k}, \omega_n)$-space. Since the largest contribution in Eq. (73) comes from energies close to the FS, we have used the renormalization parameters at the Fermi energy $Z(\mathbf{q})$ (66) and $X(\mathbf{q})$ instead of the energy dependent ones. The results for $T_c(\delta)$ is shown in the right panel in Fig. 7. The highest $T_c \sim 0.021t \sim 100$ K is found when all the contributions are taken into account, though pairing induced only by

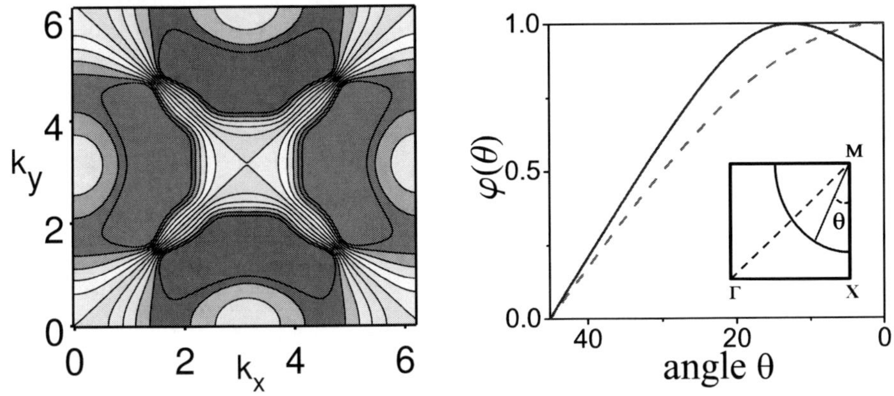

Figure 8. (Color online) Left panel: 2D plot of the SC gap $\varphi(\mathbf{k}, \omega \simeq 0)$. Right panel: Angle dependence of the SC gap $\varphi(\theta)$ on the FS (blue bold line) in comparison with the model d-wave dependence $\varphi_d(\theta) = \cos 2\theta$ (red dashed lines).

spin-fluctuations also results in high $T_c^{sf} \sim 0.014t \sim 65$ K. The d-wave pairing induced only by the EPI is rather weak and does not displayed in the right panel in Fig. 7. The value of T_c is reduced by an order of magnitude in comparison with the WCA in the left panel in Fig. 7 due to a suppression of the QP weight by the factor $[1/Z(\mathbf{q})]^2$. The maximum value of T_c is found at lower value of doping $\delta_{opt} \approx 0.12$ than in experiments, $\delta_{opt}^{exp} = 0.16$.

The $\mathbf{k}$-dependence of the gap function $\varphi(\mathbf{k}, \omega \simeq 0)$ at doping $\delta = 0.13$ for $(0 \leq k_x,\, k_y \leq 2\pi)$ is plotted in the left panel in Fig. 8. The gap reveals a distinct d-wave symmetry with maximum values in the vicinity of the FS. As shown in the right panel in Fig. 8, its angle dependence on the FS is close to the model d-wave dependence $\varphi_d(\theta) = \cos 2\theta$. Energy dependence (in units of t) of the gap function $\varphi(\mathbf{k}, \omega)$, the real and imaginary parts, is presented in Fig. 9 at $\mathbf{k} \approx (0, \pi/2)$ and $\delta = 0.13$. Since the gap function was obtained as a solution of the linear equation at $T = T_c$ the value of the gap is given in arbitrary units. The energy variation of the gap occurs in the region of $\omega \lesssim 0.4t$, of the order of the characteristic spin-fluctuation energy $\omega_s = J = 0.4t$. T_c dependence on the intersite

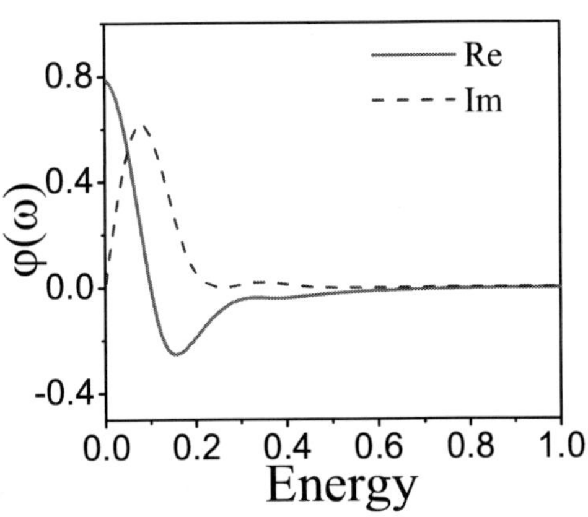

Figure 9. Energy dependence of the real, Re $\varphi(\mathbf{k}, \omega)$, and imaginary, Im $\varphi(\mathbf{k}, \omega)$, parts of the SC gap in arbitrary units.

Coulomb repulsion V is shown in the left panel in Fig. 10 for $V = 0.0, 0.5, 1.0$, and 2. Increasing of V suppresses T_c which becomes small only for high values of $V = 2t - 3t$ comparable with the spin-fluctuation coupling g_{sf} and much larger than the exchange interaction $J = 0.4t$. The maximum $T_c^{\max}$ at the optimal doping as a function of U and V is shown in the right panel in Fig. 10. It is remarkable, that $T_c^{\max}$ only weakly depends on U that supports the kinematical mechanism of pairing where the coupling constant $|t(\mathbf{q})|^2$ does not depend on U. A weak increase of $T_c^{\max}$ with U is explained by narrowing of the electronic band as seen in Figs. 1, 2 and corresponding increase of the density of state.

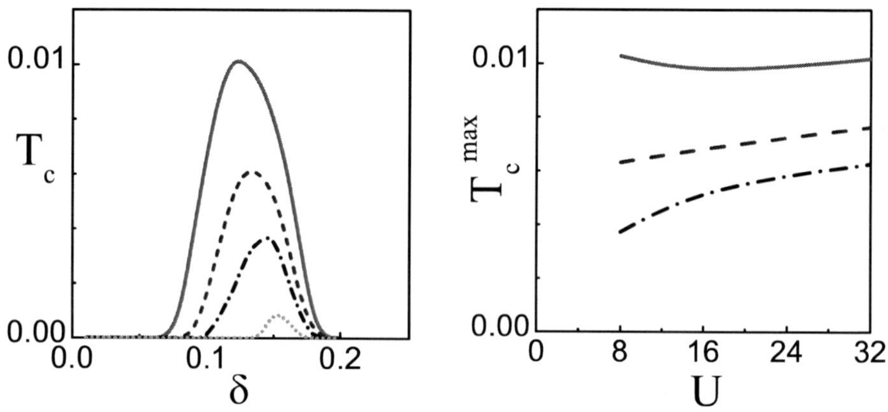

Figure 10. (Color online) Left panel: $T_c(\delta)$ for $V = 0.0$ (bold red line), $V = 0.5$ (blue dashed line), $V = 1.0$ (black dash-dotted line), and $V = 2.0$ (green dotted line) for $U = 8$. Right panel: Maximum $T_c^{\max}$ at the optimal doping as a function of U for $V = 0.0$ (bold red line), $V = 0.5$ (blue dashed line), and $V = 1.0$ (black dash-dotted line).

In the current approach one can also consider the s-wave pairing. For the extended s-wave gap function, $\varphi_s(\mathbf{k}) = (\Delta/2)\,\eta_s(\mathbf{k})$ where $\eta_s(\mathbf{k}) = (\cos k_x + \cos k_y)$, a similar to (67) equation for T_c can be derived. Solution of this equation reveals a finite and quite high T_c. However, s-wave pairing symmetry violates a kinematical restriction of no double occupancy for the Hubbard model in the two-subband regime. As was pointed out in Refs. [32, 33], the single-site correlation function should obey the condition

$$\langle X_i^{\bar{\sigma}2} X_i^{\sigma 2} \rangle = \frac{1}{N} \sum_{\mathbf{q}} \langle X_{-\mathbf{q}}^{\bar{\sigma}2} X_{\mathbf{q}}^{\sigma 2} \rangle = 0, \tag{77}$$

caused by the multiplication rule for the Hubbard operators, $X_i^{\alpha\beta} X_i^{\gamma\delta} = \delta_{\beta\gamma} X_i^{\alpha\delta}$. In the QP approximation used in Eq. (67) we obtain the relation

$$\langle X_i^{\bar{\sigma}2} X_i^{\sigma 2} \rangle = \frac{1}{N} \sum_{\mathbf{q}} \frac{\varphi(\mathbf{q})}{[Z(\mathbf{q})]^2\, 2\widetilde{\varepsilon}(\mathbf{q})} \tanh \frac{\widetilde{\varepsilon}(\mathbf{q})}{2T_c} = 0. \tag{78}$$

For the d-wave pairing $\varphi_d(\mathbf{q}) = (\Delta/2)\,(\cos q_x - \cos q_y)$ this condition is fulfilled in the tetragonal phase for any doping (pairing in the orthorhombic phase is considered in Ref. [61]). For the s-wave pairing this condition

$$\frac{1}{N} \sum_{q_x, q_y} \frac{\cos q_x}{[Z(\mathbf{q})]^2\, 2\widetilde{\varepsilon}(\mathbf{q})} \tanh \frac{\widetilde{\varepsilon}(\mathbf{q})}{2T} \neq 0, \tag{79}$$

is violated for an arbitrary doping except for a particular choice of the chemical potential when the contribution from the integral over $0 \le q_x \le \pi$ is compensated by the integral over $\pi \le q_x \le 2\pi$. The same condition holds for the one-particle subband, $\langle X_i^{0\bar{\sigma}} X_i^{0\sigma} \rangle = 0$. The obtained results can be derived for a general representation for the correlation function

$$\langle X_{-\mathbf{q}}^{\bar{\sigma}2} X_{\mathbf{q}}^{\sigma 2} \rangle = -\frac{1}{\pi Q_2 N} \sum_{\mathbf{q}} \int_{-\infty}^{+\infty} \frac{dz}{e^{z/T}+1} \mathrm{Im}\, F_{\sigma}^{22}(\mathbf{q}, z),$$

since the symmetry of the anomalous GF $F_{\sigma}^{22}(\mathbf{q}, z)$ is determined by the s- or d-wave symmetry of the gap function. Therefore, we conclude that s-wave pairing is prohibited for the Hubbard model in the limit of strong correlations.

4.4. Isotope Effect

The observation of the isotope effect (IE) in conventional superconductors, i.e., the dependence of T_c on the mass M of the lattice ions, $T_c \propto M^{-\alpha}$, with the isotope exponent $\alpha = -d \ln T_c / d \ln M \simeq 0.5$ was a direct evidence of the electron-phonon pairing mechanism in these materials. In cuprate superconductors a weak isotope shift with $\alpha \le 0.1$ was found at optimal doping. In the underdoped region isotope exponent increases and can be even larger than in conventional superconductors, $\alpha \sim 1$ (see, e.g., reviews [16, 17, 18] and references therein). Small values of the isotope exponent at optimal doping suggests nonphononic, e.g., magnetic, mechanism of pairing. However, it is difficult to explain the IE on T_c within the spin-fluctuation pairing mechanism. By taking into account the EPI within the extended Hubbard model (3) we can observe the IE in qualitative agreement with experiments.

To study the isotope effect on T_c using the gap equation (68) we consider the mass-dependent phonon frequency ω_0 in the θ-function in (68), $\omega_0 = \omega_0^{(0)} \sqrt{M_0/(M_0 + \Delta M)} = \omega_0^{(0)}(1-\beta)$. For instance, for the oxygen isotope shift $^{16}\mathrm{O} \rightarrow {}^{18}\mathrm{O}$ we have $\beta = 0.057$. We neglect the polaronic effect for the d-wave electron-phonon coupling constant $g_{ep}(\mathbf{q}) = |g(\mathbf{q})|^2 \chi_{ph}(\mathbf{q})$ assuming it to be mass-independent (for a discussion see Ref. [62]). It is easy to obtain an estimation for the isotope exponent by considering solution of the BCS-like equation (68) by taking into account only two effective coupling constant, the spin-fluctuation g_{sf} and EPI g_{ep} averaged over wave-vectors at the FS for the d-wave pairing. Then for the isotope exponent we found:

$$\alpha = -\frac{d \log T_c}{d \log M} = \frac{1}{2} \frac{g_{ep}}{(g_{ep} + g_{sf})}, \tag{80}$$

For $g_{sf} \gg g_{ep}$ we have $\alpha \approx (1/2)\,(g_{ep}/g_{sf})$, while for $g_{sf} \approx g_{ep}$ we get $\alpha \approx 1/4 = 0.25$. Similar results were obtained in Ref. [63] for the t-J model with the EPI. The result of numerical solution of the gap equation (68) for the isotope exponent is shown in Fig. 11 for two values of EPI $g_{ep} = 5\,t$ and $g_{ep} = 2.5\,t$ in Eq. (71) for $Z(\mathbf{q}) = 5$. In accordance with the analytical estimation (80), the isotope exponent increases for larger values of g_{ep}. The doping dependence of the exponent agrees with experiments, it is quite small, $\alpha = 0.09 - 0.18$, at doping close to the optimal, while drastically increases in the underdoped case, $\alpha = 0.38 - 0.68$ at $\delta = 0.1$ for $g_{ep} = 2.5\,t - 5\,t$, respectively.

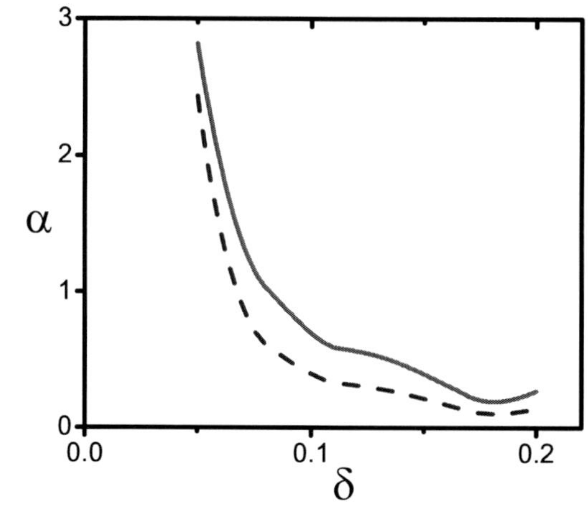

Figure 11. (Color online) Isotope exponent $\alpha(\delta)$ for $g_{ep} = 5\,t$ (red solid line) and $g_{ep} = 2.5\,t$ (blue dashed line).

4.5. Comparison with Previous Theoretical Studies

In studies of superconductivity in the framework of the conventional Hubbard model in the limit of strong correlations usually MFA was employed (see, e.g., Refs. [64, 65, 66] and references therein). As was shown in Sec 3.2, superconducting pairing in MFA is induced by the exchange interaction $J_{ij} = 4t^2/U$ (see Eq. (43)). The exchange interaction vanishes in the limit $U \to \infty$, a feature which explains the disappearance of the pairing at large U observed in Refs. [64, 65, 66]. To obtain nonzero pairing in the limit $U \gg t$ the self-energy contribution induced by the kinematical interaction should be taken into account which does not depend on U as shown in the right panel in Fig. 10. The self-energy effects in the normal state, e.g., the pseudogap formation, in the conventional Hubbard model have been studied in several papers (see, e.g., Refs. [38, 67]).

In the limit of strong correlations various numerical methods were extensively used. Here we refer to numerical simulations for finite clusters (see reviews [68, 69, 70, 71]), the DMFT (see reviews [57, 58]), the dynamical cluster approximation (DCA) [72, 73] and the cluster DMFT (see, e.g., Refs. [59, 74]). More accurate results have been obtained within the DCA and cluster DMFT methods where short-range AF correlations are partially taken into account. Extensive numerical studies for finite clusters have revealed a tendency to the d-wave pairing in the Hubbard model, though a delicate balance between superconductivity and other instabilities (AF, spin-density wave, charge-density wave, etc.) was found (see, e.g., Refs. [69, 70, 72, 73, 74]). In Ref. [75]), using the DCA with the quantum Monte Carlo method, the superconducting d-wave pairing and the isotope effect similar to observed in cuprates were found for the Hubbard-Holstein model. However, in several publications an appearance of the long-range superconducting order has not been confirmed (see, e.g., Ref. [76]).

As discussed in Sec. 1, the intersite Coulomb repulsion V is detrimental for pairing induced by the on-site CI U in the Hubbard model or higher-order contributions from V in the weak correlation limit. Here we would like to comment on several studies of this problem in the strong correlation limit and to compare them with our analytical results for the d-wave pairing. Following the original idea of Anderson [19], it is commonly believed that the exchange interaction $J = 4t^2/U$ induced by the interband hopping in the Hubbard model plays the major role in the d-wave superconducting pairing. Since the excitation energy of electrons in the interband hopping U is much larger than their intraband kinetic energy W the exchange pairing has no retardation effects contrary to the electron-phonon pairing where large Bogoliubov-Tolmachev logarithm [77] diminishes the Coulomb repulsion $V \to V/[1 + \rho_c \ln(\mu/\omega_{ph})]$ where $\rho_c = N(0)\,V$ and ω_{ph} is the phonon energy. Consequently, without the retardation effects the Coulomb repulsion V should destroy the exchange pairing for $V > J$.

To get over this problem, in Ref. [55] it was suggested that the intersite Coulomb repulsion V decreases the interband excitation energy which results in enhancement of the exchange interaction, $\widetilde{J}(V) = 4\,t^2/(U - V)$, as was found from cluster calculations. If we consider pairing induced only by the exchange interaction $\widetilde{J}(V)$ and take into account the Coulomb repulsion V then the condition $\widetilde{J}(V) - V > 0$ should be fulfilled for existence of pairing. But this condition can be satisfied only for small values of $V \ll U$. Therefore, the pure exchange superconducting pairing can occur only in the region of weak Coulomb repulsion V. Contrary to this, in Ref. [55] using the cellular dynamical mean-Field theory (CDMFT) [71] the d-wave pairing was found in the region of strong coupling up to $V \lesssim U/2$. At the same time, in the limit of weak correlations $U = 4t$ the pairing is suppressed at small value of $V \sim 1.5\,t$. Thus, we believe that the d-wave superconductivity observed by the authors is not due to renormalization of the exchange interaction $\widetilde{J}(V)$ but due to another mechanism of pairing not explicitly seen in the CDMFT calculations. As we have shown in the strong correlation limit in the two-subband regime the kinematical interaction is responsible for the spin-fluctuation pairing at large values of V, up to $V \lesssim 4\,t$ (see left panel in Fig. 10).

Our conclusion concerning importance of the kinematical mechanism of pairing is supported by the studies in Ref. [54]. Using the variational Monte Carlo technique the superconducting d-wave gap was calculated for the extended Hubbard model with a weak exchange interaction $J = 0.2\,t$ and a repulsion $V \leq 3\,t$ in a broad range of $0 \leq U \leq 32$. It was found that the gap decreases with increasing V at all U and can be suppressed for $V > J$ for small U. But for large $U \gtrsim U_c \sim 6\,t$ the gap becomes robust and exists up to large values of $V \sim 10\,J = 2\,t$ which was explained by effective enhancement of J as in Ref. [55]. At the same time, the gap does not show notable variation with U for large $U = 10 - 30$ though it should depend on the conventional exchange interaction in the Hubbard model $J = 4t^2/U$ (or $J = 4t^2/(U - V)$). We can suggest another explanation of these results by pointing out that at large $U \gtrsim U_c$ concomitant decrease of the bandwidth (as shown in Fig. 3 b) in Ref. [54]) results in the splitting of the Hubbard band into the upper and lower subbands and the emerging kinematical interaction induces the d-wave pairing in one Hubbard subband. In that case the second subband for large U gives a small contribution which results in U-independent pairing as shown in the right panel in Fig. 10. It can be suppressed by the repulsion V only larger than the kinematical interaction, $V \gtrsim 4t$.

In Ref. [78] the extended Hubbard model is considered in the weak or intermediate correlation limits as in Ref. [23] and in the strong correlation limit within the slave-boson representation in the MFA. In the strong correlation limit a small value of $V = J$ suppresses the d-wave superconducting gap. The kinetic energy term described by the projected electron operators, $t\,\hat{c}^{\dagger}_{i\sigma}\hat{c}_{j\sigma} = t\,c^{\dagger}_{i\sigma}(1-n_{i-\sigma})c_{j\sigma}(1-n_{j-\sigma}) \equiv t\,X^{\sigma 0}_{i}X^{0\sigma}_{j}$, was approximated by the conventional fermion (spinon) operators, $t\,\delta\, f^{\dagger}_{i\sigma}f_{j\sigma}$ and the most important contribution from the kinematical interaction was lost in the resulting BCS-type gap equation (13) in Ref. [78]. As shown in Eq. (68) for the gap the kinematical interaction given by $\widehat{\chi}_{sf}$ (72) provides strong spin-fluctuation pairing and high T_c. Thus, the slave-boson theory treated in MFA fails to describe superconducting pairing in the limit of strong correlations.

5. Conclusion

In our work we present the theory of superconducting pairing within the extended Hubbard model (3) in the limit of strong electron correlations. Using the Mori-type projection technique we obtained a self-consistent system of equations for normal and anomalous (pair) GFs and for the self-energy calculated in the SCBA. The theory is similar to the Migdal-Eliasberg strong-coupling theory.

From our study we can draw the following conclusions concerning the mechanism of pairing in the extended Hubbard model. Solution of the gap equation (67) shows that for the d-wave pairing the intersite Coulomb repulsion gives a small contribution determined by the angular momentum $l = 2$ of the interaction function $V(\mathbf{k}-\mathbf{q})$. However, it can be larger than the AF superexchange interaction $J(\mathbf{k}-\mathbf{q})$, and the RVB-type superconducting pairing can be destroyed. Pairing induced by charge fluctuations $\chi_{ch}(\mathbf{k}-\mathbf{q})$ appears to be quite weak (outside the charge-instability region). In Sec. 4.3 we have found that the d-wave component of the EPI, even for the model with strong interaction [41] and a large fully symmetric s-wave component, turned out to be small. The largest contribution to the d-wave pairing comes from the electron coupling to spin fluctuations induced by strong kinematical interaction $|t(\mathbf{q})|^2$. Thus we conclude that the EPI plays a secondary role in achieving high-T_c though it explains the isotope effect as discussed in Sec. 4.4

It is important to point out that the superconducting pairing induced by the AF superexchange interaction and spin-fluctuation scattering are caused by the kinematical interaction characteristic of systems with strong correlations. These mechanisms of superconducting pairing are absent in the fermionic models, while they are generic for cuprates. The intersite Coulomb repulsion is not strong enough to destroy the d-wave pairing mediated by spin fluctuations. Therefore, we believe that the magnetic mechanism of superconducting pairing in the Hubbard model in the limit of strong correlations is a relevant mechanism of high-temperature superconductivity in the copper-oxide materials.

6. Appendix

6.1. Pair Correlation Function in GMFA

The pairing in the GMFA is determined by the pair correlation function $\langle X^{02}_{i} N_j\rangle$ as given by Eqs. (40), (41). In Refs. [64, 66] for its calculation the Roth procedure was used where the Hubbard operators were decoupled on the same lattice site by writing the time-dependent correlation function in the form: $\langle c_{i\downarrow}(t)|c_{i\uparrow}(t')N_j(t')\rangle = \langle X^{0\downarrow}_{i}(t)|X^{\downarrow 2}_{i}(t')N_j(t')\rangle$. In view of the identity satisfied by the Hubbard operators:

$X_i^{\alpha\beta} = X_i^{\alpha\gamma} X_i^{\gamma\beta}$ for any intermediate state $|\gamma\rangle$, the decoupling of the operators at the same lattice site is not unique and therefore unreliable (a feature already noticed in Refs. [64, 66]).

Here we perform a direct calculation of the correlation function $\langle X_i^{02} N_j \rangle$ *without any decoupling* by considering the equation for the corresponding commutator GF $L_{ij}(t-t') = \langle\langle X_i^{02}(t) \mid N_j(t') \rangle\rangle|_{i \neq j}$. The equation for the GF can be written as [37]:

$$(\omega - \varepsilon_2) L_{ij}(\omega) = \sum_{m \neq i, \sigma'} \sigma' t_{im}^{12} \{ \langle\langle X_i^{0\bar{\sigma}'} X_m^{0\sigma'} | N_j \rangle\rangle_\omega - \langle\langle X_i^{\sigma' 2} X_m^{\bar{\sigma}' 2} | N_j \rangle\rangle_\omega \}, \tag{81}$$

where we neglected excitation energy proportional to the intraband hopping in comparison with the interband contribution, $|t_{im}^{\alpha\alpha}| \ll |\varepsilon_2| \simeq U$. The pair correlation function is determined by the equation:

$$\langle X_i^{02} N_j \rangle = -\frac{1}{\pi} \int_{-\infty}^{+\infty} \frac{d\omega}{1 - \exp(-\omega/T)} \operatorname{Im} L_{ij}(\omega). \tag{82}$$

The GF $L_{ij}(\omega)$ has two poles, one at $\omega = \varepsilon_2$ and another at the energy of a pair excitation given by the GFs at the right-hand side in Eq. (81) of the singly or the doubly occupied subbands. Let us consider the hole doped case, $n = 1 + \delta > 1$ when the chemical potential crosses the two-hole subband, $\mu \sim U$ and $\varepsilon_1 = -\mu \sim -U$, $\varepsilon_2 = 2\varepsilon_1 + U \sim -U$. In this case we can neglect the exponentially small contribution of the order of $\exp(-U/T) \ll 1$ coming from the pole $\omega = \varepsilon_2$. The contribution from the GF of the one-hole subband in (81),

$$-\frac{1}{\pi} \operatorname{Im} \langle\langle X_i^{0\bar{\sigma}} X_m^{0\sigma} | N_j \rangle\rangle_\omega \simeq \delta_{mj} \langle X_i^{0\bar{\sigma}} X_j^{0\sigma} \rangle \delta(\omega - 2\varepsilon_1), \tag{83}$$

also gives an exponentially small contribution of the order of $\exp(-2U/T) \ll 1$. Therefore, we can take into account only the contribution from the GF $\langle\langle X_i^{\sigma' 2} X_m^{\bar{\sigma}' 2} | N_j \rangle\rangle_\omega$ where the pair excitation energy $\omega \sim |t_{im}^{22}|$. Using the approximation $1/(\omega - \varepsilon_2) \simeq 1/U$ which neglects retardation effects an integration over ω in Eq. (82) gives the following result,

$$\begin{aligned} \langle X_i^{02} N_j \rangle &= -\frac{1}{U} \sum_{m \neq i, \sigma'} \sigma' t_{im}^{12} \langle X_i^{\sigma' 2} X_m^{\bar{\sigma}' 2} N_j \rangle \\ &\simeq -(4 t_{ij}^{12}/U) \sigma \langle X_i^{\sigma 2} X_j^{\bar{\sigma} 2} \rangle. \end{aligned} \tag{84}$$

The last formula is obtained in the two-site approximation usually applied for the t-J model: $m = j$, which gives $X_j^{\bar{\sigma}2} N_j = 2 X_j^{\bar{\sigma}2}$.

6.2. Self-Energy

The normal and anomalous (pair) components of the self-energy operator (29) are given by the matrices:

$$\hat{M}_{ij\sigma}(\omega) = \langle\langle \begin{pmatrix} \widetilde{[X_i^{\sigma 2}, H]} \\ \widetilde{[X_i^{0\bar{\sigma}}, H]} \end{pmatrix} \mid (\widetilde{[H, X_j^{2\sigma}]} \; \widetilde{[H, X_j^{\bar{\sigma}0}]}) \rangle\rangle_\omega^{(pp)}, \tag{85}$$

$$\hat{\Phi}_{ij\sigma}(\omega) = \langle\langle \begin{pmatrix} \widetilde{X_i^{\sigma 2}, H]} \\ \widetilde{[X_i^{0\bar{\sigma}}, H]} \end{pmatrix} \mid (\widetilde{[H, X_j^{\bar{\sigma} 2}]} \widetilde{[H, X_j^{0\sigma}]}) \rangle\rangle_\omega^{(pp)}, \tag{86}$$

where $\widetilde{[X_i^{\alpha,\beta}, H]}$ and $\widetilde{[H, X_i^{\alpha,\beta}]}$ are the irreducible parts of the commutators determined by Eq. (21). Using equations of motion for the HOs as given, e.g., by Eq. (11) we obtain multiparticle GFs which are determined by products of bosonic and fermionic operators. Let us consider, in particular, contributions to the two-hole subband self-energy given by the kinematical interaction in Eq. (11). The normal component in Eq. (85) reads,

$$\begin{aligned} M_{ij\sigma}^{22(k)}(\omega) &= \sum_{ll'\sigma'\sigma''} t_{il}t_{jl'} \langle\langle B_{i\sigma\sigma'} X_l^{\sigma'2} | X_{l'}^{2\sigma''} B_{j\sigma\sigma''}^\dagger \rangle\rangle_\omega \\ &= \sum_{ll'\sigma'\sigma''} t_{il}t_{jl'} \frac{1}{2\pi} \int_{-\infty}^{\infty} dz \frac{e^{\beta z}+1}{(\omega - z)} \int_{-\infty}^{\infty} dt e^{izt} \\ &\times \langle X_{l'}^{2\sigma''} B_{j\sigma\sigma''}^\dagger | B_{i\sigma\sigma'}(t) X_l^{\sigma'2}(t) \rangle. \end{aligned} \tag{87}$$

For the anomalous component in Eq. (86) we have

$$\begin{aligned} \Phi_{ij\sigma}^{22(k)}(\omega) &= -\sum_{ll'\sigma'\sigma''} t_{il}t_{jl'} \langle\langle B_{i\sigma\sigma'} X_l^{\sigma'2} | X_{l'}^{\bar{\sigma}''2} B_{j\bar{\sigma}\bar{\sigma}''} \rangle\rangle_\omega \\ &= -\sum_{ll'\sigma'\sigma''} t_{il}t_{jl'} \frac{1}{2\pi} \int_{-\infty}^{\infty} dz \frac{e^{\beta z}+1}{\omega - z} \int_{-\infty}^{\infty} dt e^{izt} \\ &\times \langle X_{l'}^{\bar{\sigma}''2} B_{j\bar{\sigma}\bar{\sigma}''} | B_{i\sigma\sigma'}(t) X_l^{\sigma'2}(t) \rangle, \end{aligned} \tag{88}$$

where the bosonic operator $B_{i\sigma\sigma'} = B_{i\sigma\sigma'}^{22}$ is defined by Eq. (12). The multiparticle time-dependent correlation functions are calculated in the SCBA as described by Eqs. (44), (45). The time-dependent single-particle fermionic and bosonic correlation functions which appear after the two-time decoupling are calculated self-consistently as e.g.,

$$\langle X_{l'}^{2\sigma} X_l^{\sigma 2}(t) \rangle = \int_{-\infty}^{\infty} d\omega' n(\omega') e^{-i\omega' t} [-(1/\pi)] \, \mathrm{Im} G_{ll'\sigma}^{22}(\omega'), \tag{89}$$

$$\langle B_{j\sigma\sigma'}^\dagger | B_{i\sigma\sigma'}(t) \rangle = \int_{-\infty}^{\infty} d\omega' N(\omega') e^{-i\omega' t} [-(1/\pi)] \, \mathrm{Im} \langle\langle B_{i\sigma\sigma'} | B_{j\sigma\sigma'}^\dagger \rangle\rangle_{\omega'}. \tag{90}$$

Here $G_{ll'\sigma}^{22}(\omega')$ is the GF (30) for the two-hole subband and $\langle\langle B_{i\sigma\sigma'} | B_{j\sigma\sigma'}^\dagger \rangle\rangle_{\omega'}$ is the commutator GF for bosonic excitations. Integration over time t in Eqs. (87) and (88) yields

$$\begin{aligned} M_{ij\sigma}^{22(k)}(\omega) &= \iint_{-\infty}^{\infty} \frac{d\omega_1 d\omega_2}{\pi^2} \frac{1 - n(\omega_1) + N(\omega_2)}{\omega - \omega_1 - \omega_2} \\ &\times \sum_{ll'\sigma'} t_{il}t_{jl'} \, \mathrm{Im} G_{ll'\sigma'}^{22}(\omega_1) \, \mathrm{Im} \langle\langle B_{i\sigma\sigma'} | B_{j\sigma\sigma'}^\dagger \rangle\rangle_{\omega_2}, \end{aligned} \tag{91}$$

$$\begin{aligned} \Phi_{ij\sigma}^{22(k)}(\omega) &= -\iint_{-\infty}^{\infty} \frac{d\omega_1 d\omega_2}{\pi^2} \frac{1 - n(\omega_1) + N(\omega_2)}{\omega - \omega_1 - \omega_2} \\ &\times \sum_{ll'\sigma'} t_{il}t_{jl'} \, \mathrm{Im} F_{ll'\sigma'}^{22}(\omega_1) \, \mathrm{Im} \langle\langle B_{i\sigma\sigma'} | B_{j\bar{\sigma}\bar{\sigma}'} \rangle\rangle_{\omega_2}. \end{aligned} \tag{92}$$

Taking into account the definition of the bosonic operator (12) the bosonic GFs in these equations can be written as

$$\langle\langle B_{i\sigma\sigma'} | B_{j\sigma\sigma'}^\dagger \rangle\rangle_\omega = (1/4) \langle\langle N_i | N_j \rangle\rangle_\omega \, \delta_{\sigma'\sigma}$$

$$+\langle\langle S_i^z|S_j^z\rangle\rangle_\omega\,\delta_{\sigma'\sigma} + \langle\langle X_i^{\bar{\sigma}\sigma}|X_j^{\sigma\bar{\sigma}}\rangle\rangle_\omega\,\delta_{\sigma'\bar{\sigma}}, \tag{93}$$

$$\langle\langle B_{i\sigma\sigma'}|B_{j\bar{\sigma}\bar{\sigma}'}\rangle\rangle_\omega = (1/4)\langle\langle N_i|N_j\rangle\rangle_\omega\,\delta_{\sigma'\sigma}$$
$$-\langle\langle S_i^z|S_j^z\rangle\rangle_\omega\,\delta_{\sigma'\sigma} + \langle\langle X_i^{\bar{\sigma}\sigma}|X_j^{\sigma\bar{\sigma}}\rangle\rangle_\omega\,\delta_{\sigma'\bar{\sigma}}. \tag{94}$$

After summation over σ' in (91) for the bosonic GF (93) and the normal GF in the paramagnetic state, $G^{22}_{ll'\sigma}(\omega) = G^{22}_{ll'\bar{\sigma}}(\omega)$, the spin-fluctuation contribution can be written in the form: $\langle\langle S_i^z|S_j^z\rangle\rangle_\omega + \langle\langle X_i^{\bar{\sigma}\sigma}|X_j^{\sigma\bar{\sigma}}\rangle\rangle_\omega = \langle\langle \mathbf{S}_i|\mathbf{S}_j\rangle\rangle_\omega$. Similar summation over σ' in (92) for the bosonic GF (94) and the anomalous GF $F^{22}_{ll'\sigma}(\omega) = -F^{22}_{ll'\bar{\sigma}}(\omega)$, results in the equation: $-\langle\langle S_i^z|S_j^z\rangle\rangle_\omega\, F^{22}_{ll'\sigma}(\omega) + \langle\langle X_i^{\bar{\sigma}\sigma}|X_j^{\sigma\bar{\sigma}}\rangle\rangle_\omega\, F^{22}_{ll'\bar{\sigma}}(\omega) = -\langle\langle \mathbf{S}_i|\mathbf{S}_j\rangle\rangle_\omega\, F^{22}_{ll'\sigma}(\omega)$.

Introducing the **q**-representation for the GFs and the self-energies as defined by Eq. (16) for the self-energies (91) and (92) we obtain the expressions:

$$M^{22(k)}(\mathbf{k},\omega) = \frac{1}{N}\sum_{\mathbf{q}}\int_{-\infty}^{+\infty} \mathrm{d}z\, K^{(+)}_{(k)}(\omega, z|\mathbf{q},\mathbf{k}-\mathbf{q})$$
$$\times\ [-(1/\pi)\,\mathrm{Im}\,G^{22}(\mathbf{q},z)], \tag{95}$$

$$\Phi^{22(k)}_{\sigma}(\mathbf{k},\omega) = \frac{1}{N}\sum_{\mathbf{q}}\int_{-\infty}^{+\infty} \mathrm{d}z\, K^{(-)}_{(k)}(\omega, z|\mathbf{q},\mathbf{k}-\mathbf{q})$$
$$\times\ [-(1/\pi)\,\mathrm{Im}\,F^{22}_{\sigma}(\mathbf{q},z)], \tag{96}$$

where the contribution from the kinematical interaction is given by the kernel

$$K^{(\pm)}_{(k)}(\omega, z|\mathbf{q},\mathbf{k}-\mathbf{q}) = \frac{|t(\mathbf{q})|^2}{\pi}\int_{-\infty}^{+\infty} d\Omega \frac{1+N(\Omega)-n(z)}{\omega - z - \Omega}$$
$$\times\{\mathrm{Im}\,\chi_{sf}(\mathbf{k}-\mathbf{q},\Omega) \pm (1/4)\mathrm{Im}\,\chi_{cf}(\mathbf{k}-\mathbf{q},\Omega)\}. \tag{97}$$

Here the spin- and charge-susceptibility are defined by Eqs. (49) and (50). By taking into account contributions from the CI and the EPI in Eq.(11) and using the SCBA in calculation of the respective time-dependent correlation functions we obtain the kernel (48) for the integral equations (46) and (47).

References

[1] J. G. Bednorz and K. A. Müller, *Z. Phys. B.* 64, 189 (1986).

[2] *Handbook of High-Temperature Superconductivity. Theory and Experiment*, edited by J.R. Schrieffer and J.S. Brooks (Springer-Verlag, New York, 2007), Chaps. 13–15.

[3] N. M. Plakida, *High-Temperature Cuprate Superconductors* (Springer Series in Solid-State Sciences, Vol. 166, Springer-Verlag, Berlin, 2010), Chap. 7.

[4] M. L. Kulić, in *Lectures on Physics of Highly Correlated Electronic Systems VIII*, edited by A. Avella and F. Mancini, *AIP Conf. Proc.*, Vol. 715 (Melville, New York, 2004), p. 75.

[5] E.G. Maksimov, M.L. Kulić, and O. V. Dolgov, *Adv. in Cond. Mat. Phys.*, Volume 2010, Article ID 423725 (2010) (DOI: 10.1155/2010/423725).

[6] D.J. Scalapino, *Phys. Reports* 250, 329 (1995).

[7] P. Monthoux and D. Pines, *Phys. Rev. B* 49, 4261 (1994).

[8] T. Moriya and K. Ueda, *Adv. in Physics* 49, 555 (2000); *Rep. Prog. Phys.* 66, 1299 (2003).

[9] A. V. Chubukov, D. Pines, and J. Schmalian, in *The Physics of Conventional and Unconventional Superconductors*, edited by K. H. Bennemann and J. B. Ketterson (Springer-Verlag, Berlin, 2004), Vol. I, p. 495; Ar. Abanov, A.V. Chubukov, and J. Schmalian, *Advances in Phys.* 52, 119 (2003).

[10] Ar. Abanov, A.V. Chubukov, and M.R. Norman, *Phys. Rev. B* 78, 220507(R) (2008).

[11] Ph. Bourges, in *The Gap Symmetry and Fluctuations in High Temperature Superconductors*, edited by J. Bok, G. Deutscher, D. Pavuna and S. A. Wolf (Plenum Press, 1998), p. 349.

[12] M. Le Tacon, G. Ghiringhelli, J. Chaloupka, M. Moretti Sala, V. Hinkov, M.W. Haverkort, M. Minola, M. Bakr, K. J. Zhou, S. Blanco-Canosa, C. Monney, Y. T. Song, G. L. Sun, C. T. Lin, G. M. De Luca, M. Salluzzo, G. Khaliullin, T. Schmitt, L. Braicovich, and B. Keimer, *Nature Phys.* 7, 725 (2011).

[13] M. Le Tacon, M. Minola, D. C. Peets, M. Moretti Sala, S. Blanco-Canosa, V. Hinkov, R. Liang, D. A. Bonn, W. N. Hardy, C. T. Lin, T. Schmitt, L. Braicovich, G. Ghiringhelli, and B. Keimer, *Phys. Rev. B* 88, 020501(R) (2013).

[14] A.A. Kordyuk, V.B. Zabolotnyy, D.V. Evtushinsky, D.S. Inosov, T.K. Kim, B. Büchner, and S.V. Borisenko, *Eur. Phys. J.* 188, 153 (2010).

[15] A.A. Vladimirov, D. Ihle, and N.M. Plakida, *Phys. Rev. B* 85, 224536 (2012).

[16] G.-M. Zhao, H. Keller, and K. Conder, *J. Phys.: Condens. Matter* 13, R569 (2001).

[17] R. Khasanov, A. Shengelaya, E. Morenzoni, K. Conder, I.M. Savič, and H. Keller, *J. Phys.: Condens. Matter* 16, S4439 (2004).

[18] K.A. Müller, in *Handbook of High-Temperature Superconductivity. Theory and Experiment*, ed. by J.R. Schrieffer and J.S. Brooks (Springer, New York, 2007), pp. 1–18.

[19] P. W. Anderson, *Science* 235, 1196 (1987); P. W. Anderson, *The theory of superconductivity in the high-T_c cuprates* (Princeton University Press, Princeton, 1997).

[20] J. Hubbard, *Proc. Roy. Soc.* (London) A, 276, 238 (1963).

[21] S. Raghu, S. A. Kivelson, and D.J. Scalapino, *Phys. Rev. B* 81, 224505 (2010).

[22] A. S. Alexandrov and V.V. Kabanov, *Phys. Rev. Lett.* 106, 136403 (2011).

[23] S. Raghu, E. Berg, A.V. Chubukov, and S. A. Kivelson1 *Phys. Rev. B* 85, 024516 (2012).

[24] W. Kohn and J. M. Luttinger, *Phys. Rev. Lett.* 15, 524 (1965).

[25] M. Yu. Kagan, D.V. Efremov, M.S. Marienko, and V.S. Valkov. *JETP Lett.* 93, 725 (2011).

[26] D. V. Efremov, M. S. Marenko, M.A. Baranov, and M.Yu. Kagan, *J. Exp. Theor. Phys.* 90, 861 (2000).

[27] M. Yu. Kagan, V. V. Val'kov, V. A. Mitskan, and M. M. Korovushkin, *J. Exp. Theor. Phys.* 144, 837 (2013).

[28] P. Fulde, *Electronic correlations in molecules and solids* (Springer Verlag, Berlin 1995).

[29] *Strongly Correlated Systems. Theoretical Methods*, edited by A. Avella and F. Mancini (Springer Series in Solid-State Sciences, Vol. 171, Springer Verlag, Berlin, 2012).

[30] J. Hubbard, *Proc. Roy. Soc. A* (London), 285, 542 (1965).

[31] R.O. Zaitsev and V.A. Ivanov, *Soviet Phys. Solid State* 29, 2554 (1987), *ibid.* 29, 3111 (1987), *Int. J. Mod. Phys. B* 5, 153 (1988).

[32] N.M. Plakida, V.Yu. Yushankhai, and I.V. Stasyuk, *Physica C* 160, 80 (1989).

[33] V.Yu. Yushankhai, N.M. Plakida, and P. Kalinay, *Physica C* 174, 401 (1991).

[34] N.M. Plakida and V.S. Oudovenko, *Phys. Rev. B* 59, 11949 (1999).

[35] P. Prelovšek and A. Ramšak, *Phys. Rev. B* 72, 012510 (2005).

[36] N.M. Plakida, *Physica C* 282–287, 1737 (1997).

[37] N.M. Plakida, L. Anton, S. Adam, and Gh. Adam, *Zh. Exp. Theor. Fyz.* 124, 367 (2003), (*JETP* 97, 331 (2003)).

[38] N.M. Plakida and V.S. Oudovenko, *JETP* 104, 230 (2007).

[39] N. M. Plakida and V. S. Oudovenko, *Eur. Phys. J. B* 86, 115 (2013).

[40] N. M. Plakida and V. S. Oudovenko, *Zh. Exp. Theor. Fyz.*, 146, 631(2014) [*JETP* 119, 554 (2014)].

[41] R. Zeyher and M. L. Kulić, *Phys. Rev. B* 53, 2850 (1996).

[42] H. Mori, *Prog. Theor. Phys.* 34, 399 (1965).

[43] N. M. Plakida, Projection operator method, in: *Theoretical Methods for Strongly Correlated Systems*, Eds. A. Avella and F. Mancini (Springer Series in *Solid-State Sciences* Vol. 171, Springer, Heidelberg, 2011) Chap. 6, pp. 173–202.

[44] D.N. Zubarev, *Usp. Fiz. Nauk* 71, 71 (1960); (*Sov. Phys. Usp.* 3, 320 (1960)); *Nonequilibrium Statical Thermodynamics* (Consultant Bureau, New-York, 1974).

[45] N.N. Bogoliubov, *Lektsii z kvantovoi statistiki. Pitannia statistichnoi mekhaniki kvantovykh system*, (in Ukranien). (Radz. Shkola, Kiiev, 1949) p. 227. English translation: *Lectures on Quantum Statistics* (Gordon and Breach, Sci. Publ., Inc., New York, 1967).- Vol. 1: *Quantum Statistics*, p. 250.

[46] F. Dyson, *Phys. Rev.* 102, 1217 (1956).

[47] Gh. Adam and S. Adam, *J. Phys. A: Math. Theor.* 40, 11205 (2007).

[48] Yu. A. Izyumov and B. M. Letfulov, *J. Phys.: Condens. Matter* 3 5373 (1991), *Intern. J. Modern Phys. B* 6, 321 (1992).

[49] A.B. Migdal, Zh. Eksp. *Teor. Fiz.* 34, 1438 (1956), [*Soviet Phys. JETP* 7, 996 (1958)].

[50] G.M. Eliashberg, *Zh. Eksp. Teor. Fiz.* 38, 966 (1960); *ibid* 39, 1437 (1960) [*Soviet Phys. JETP* 11, 696 (1960); ibid 12, 1000 (1960)].

[51] A.A. Vladimirov, D. Ihle, and N. M. Plakida, *Phys. Rev. B* 80, 104425 (2009).

[52] J. Jaklič and P. Prelovśek, *Phys. Rev. Lett.* 74, 3411 (1995); ibid. **75**, 1340 (1995).

[53] A.I. Lichtenstein. and M.L. Kulić, *Physica C* 245, 186 (1995).

[54] E. Plekhanov, S. Sorella, and M. Fabrizio, *Phys. Rev. Lett.* 90, 187004 (2003).

[55] D. Sénéchal, A. Day, V. Bouliane, and A.-M. S. Tremblay, arXiv:1212.4503.

[56] A.A. Kordyuk, S.V. Borisenko, A. Koitzsch, J. Fink, M. Knupfer, and H. Berger, *Phys. Rev. B* 71, 214513 (2005).

[57] A. Georges, G. Kotliar, W. Krauth, and M. Rozenberg, *Rev. Mod. Phys.* 68, 13 (1996).

[58] G. Kotliar, S. Y. Savrasov, K. Haule, V.S. Oudovenko, O. Parcollet, and C.A. Marianetti, *Rev. Mod. Phys.* 78, 865 (2006).

[59] K. Haule and G. Kotliar, *Phys. Rev. B* 76, 104509 (2007).

[60] K.M. Shen, F. Ronning, Lu, D.H., Baumberger, F., Ingle, N.J.C., Lee, W.S., Meevasana, W., Kohsaka, Y., Azuma, M., Takano, M., Takagi, H., and Z.-X. Shen, *Science* 307, 901 (2005).

[61] N. M. Plakida and V. S. Oudovenko, *Physica C*, 341–348, 289 (2000); in *Proceedings of the NATO ARW on open problems in strongly correlated systems*, edited by J.Bonca, P. Prelovsek, A. Ramsak and S. Sarkar (Kluewer Academic Publs., 2001) p. 111-116.

[62] N. M. Plakida, *Phys. Scr.* 83, 038303 (2011).

[63] E.I. Shneyder and S.G. Ovchinnikov, *Zh. Eksp. Teor. Fiz.* 136, 1177 (2009).

[64] J. Beenen and D.M. Edwards, *Phys. Rev. B*, 52, 13636 (1995).

[65] A. Avella, F. Mancini, D. Villani, and H. Matsumoto, *Physica C*, 282–287, 1757 (1997); T. Di Matteo, F. Mancini, H. Matsumoto, and V.S. Oudovenko, *Physica B*, 230–232, 915 (1997).

[66] T.D. Stanescu, I. Martin, and Ph. Phillips, *Phys. Rev. B* 62, 4300 (2000).

[67] A. Avella and F. Mancini, *Phys. Rev. B* 75, 134518 (2007); A. Avella and F. Mancini, J. Phys.: *Condens. Matter* 19, 255209 (2007).

[68] E. Dagotto, *Rev. Mod. Phys.* 66, 763 (1994).

[69] N. Bulut, *Advances in Physics* 51, 1587 (2002).

[70] D.J. Scalapino, in *Handbook of High-Temperature Superconductivity. Theory and Experiment*, edited by J.R. Schrieffer and J.S. Brooks (Springer-Verlag, New York, 2007), pp. 495–526.

[71] D. Sénéchal, Cluster dynamical mean field theory, in *Theoretical methods for Strongly Correlated Systems,* edited by A. Avella and F. Mancini (Springer Series in Solid-State Sciences, Vol. 171, Springer- Verlag, Berlin, 2012), Chap. 11.

[72] Th. Maier, M. Jarrel, Th. Pruschke, and M.H. Hettler, *Rev. Mod. Phys.* 77, 1027 (2005).

[73] Th.A. Maier, M. Jarrell, and D.J. Scalapino, *Phys. Rev. Lett.* 96, 047005 (2006); ibid, *Phys. Rev. B* 74, 094513 (2006).

[74] S. S. Kancharla, B. Kyung, D. Sénéchal, M. Civelli, M. Capone, G. Kotliar, and A.-M. S. Tremblay, *Phys. Rev. B* 77, 184516 (2008).

[75] A. Macridin and M. Jarrell, *Phys. Rev. B* 79, 104517 (2009).

[76] T. Aimi and M. Imada, *J. Phys. Soc. Jpn.* 76, 13708 (2007).

[77] N. N. Bogoliubov, V. V. Tolmachev, and D. V. Shirkov, *New method in the theory of superconductivity*, (Publ. Dept. USSR Acad. of Science, Moscow, 1958; Consultants Bureau, Chapman and Hall, New York - London, 1959, Vol. YII).

[78] S. Raghu, R. Thomale, and T. H. Geballe, *Phys. Rev. B* 86, 094506 (2012).

INDEX

A

B

C

D

E

F

G

H

I

J

K

L

M

N

O

P

Q

R

S

T

U

V

W

X

Y